Outstanding Contributions to Logic

Volume 14

More information about this series at http://www.springer.com/series/10033

Melvin Fitting · Brian Rayman
Editors

Raymond Smullyan on Self Reference

Springer

Editors
Melvin Fitting
City University of New York (CUNY) System
City University of New York
New York City, NY
USA

Brian Rayman
NYC Department of Education
Brooklyn, NY
USA

ISSN 2211-2758 ISSN 2211-2766 (electronic)
Outstanding Contributions to Logic
ISBN 978-3-319-68731-5 ISBN 978-3-319-68732-2 (eBook)
https://doi.org/10.1007/978-3-319-68732-2

Library of Congress Control Number: 2017956336

Cover illustration: Melvin Fitting

Printed on acid-free paper

This Springer imprint is published by Springer Nature
The registered company is Springer International Publishing AG
The registered company address is: Gewerbestrasse 11, 6330 Cham, Switzerland

Preface

This volume began as a tribute to Raymond Smullyan, with the additional hope that it would please him. It ended as a tribute to his memory. Along the way, it did please him, in a pre-publication version.

Raymond Smullyan had a lifelong interest in self-reference. Actually, this is a considerable understatement. He returned to the subject over and over. Self-reference and fixpoints are closely related, after all. His first technical publications, preceding his Ph.D., had to do with self-reference, as did many of his technical books. Outside of mathematics, he is best-known for his popular puzzle books, and self-reference is a fundamental theme of many of these. The present volume is not a detailed formal study of Smullyan's work. Rather it consists of commentary on and extensions of this work, by colleagues, friends, and admirers. We hope it provides some enlightenment and some entertainment, which is what Raymond would have hoped for too.

New York City, USA — Melvin Fitting
Brooklyn, USA — Brian Rayman

Contents

Chapter 1
Introduction

Melvin Fitting

Raymond Smullyan was born in 1919, after the Great War and before the Great Depression. I have known him since I was a graduate student and he was my advisor. I got my PhD in 1968, so it is clear that it has been a long time that I have known him. In this introduction I will call him "Raymond." This may sound formal to those who know him as "Ray," but it is what I have always called him, and I don't feel like changing now.

Raymond has had two long, different, but overlapping careers: one as an academic mathematician and philosopher, and another as a more popular figure.[1] Mathematics and philosophy are intertwined, each influences the other, and it is essentially impossible to discuss one without the other. Further, large parts of both relate directly to our chosen topic of self-reference. We begin with a broad overview, followed by sections on his technical and on his more popular work.

1.1 The Broad Background

As an academic, Raymond has written several books and numerous research papers that have been influential, often in unexpected ways. His first book, *Theory of Formal Systems* (Smullyan 1961) was a novel presentation of recursion theory and the Gödel incompleteness theorems, topics he revisited often throughout his career. The book introduced ideas that were influential in automata theory, as well as a decidedly abstract approach to incompleteness and undecidability results concerning the foundations of mathematics. Remarkably and unexpectedly, it also anticipated some of

[1] I omit discussion of his lives as pianist and as magician.

M. Fitting (✉)
City University of New York (emeritus), Montrose, NY 10548, USA
e-mail: melvin.fitting@gmail.com

M. Fitting and B. Rayman (eds.), *Raymond Smullyan on Self Reference*,
Outstanding Contributions to Logic 14, https://doi.org/10.1007/978-3-319-68732-2_1

the ideas underlying the computer language Prolog, developed much later for use in artificial intelligence. Neither at that time nor today has Raymond claimed to know the first thing about computer science, but of course that is not a prerequisite for having an influence.

Raymond's second book, *First-Order Logic* (Smullyan 1968) was intended to be an elegant and beautiful presentation of formal logic.[2] It was indeed elegant, and influenced several generations of logicians. A few years ago it was reprinted in the Dover book series, and continues its influence. But starting in the 1980 s it has also had an unanticipated life in computer science, in the field of automated theorem proving. There is a long-running annual conference, Tableaux, with methods and programs covering many areas of logic and mathematics, and many a paper is, in a way, a direct or indirect descendent of Raymond's 1968 book. *First-Order Logic*, along with a journal paper of Raymond, *A Unifying principle in quantification theory*, (Smullyan 1965), introduced what became called the Model Existence Theorem, which later became a fundamental tool in infinitary logic, again an unexpected outcome of Raymond's search for elegance. But the areas covered in *First-Order Logic* do not involve self-reference, and we will not discuss it further here. There have been several other of Raymond's books devoted to his research in mathematical logic: *Gödel's Incompleteness Theorems* (Smullyan 1992a); *Recursion Theory for Metamathematics* (Smullyan 1993) *Diagonalization and Self-Reference* (Smullyan 1994); and jointly with Melvin Fitting, *Set Theory and the Continuum Problem* (Smullyan and Fitting 1996). All but the last of these do have self-reference as a central topic. In addition to the books mentioned above, there have been numerous research papers, a few of which will be discussed.

In the 1970s, even while continuing his formal mathematical logic researches, Raymond's career took a rather unexpected turn. He developed an interest in puzzles, especially those that were based on mathematical logic. Martin Gardner devoted a column in *Scientific American* to some of these puzzles, and they turned out to be quite popular. This led to *What is the Name of This Book?* (Smullyan 1978). It turned out to be unexpectedly successful, and was followed by a very large number of others: *This Book Needs No Title* (Smullyan 1980); *Alice in Puzzle-Land* (Smullyan 1982a), and many others. I have lost track of foreign translations—there have been a very large number. As a result of the popularity of these books, Raymond appeared on the Tonight Show with Johnny Carson, and on William F. Buckley's firing line, and other television shows as well.

At some point Raymond decided that puzzles could be used to teach some of the fundamental discoveries of modern logic. This idea was first implemented in his book *The Lady or the Tiger?* (Smullyan 1982c), which actually led readers through the basic ideas of Gödel's incompleteness theorem, via a series of puzzles. That was followed by *To Mock a Mockingbird* (Smullyan 1985), which explored the Lambda-calculus via puzzles about birds.[3] The Lambda calculus is a system of formal

[2] *Elegant* is the word most often used about Raymond's work in reviews.

[3] Raymond also explored the topic more technically in his book *Diagonalization and Self-Reference*, (Smullyan 1994).

logic that has had applications to the design and semantics of computer languages. Raymond's book was of such interest that the automated theorem-proving group at Argonne National Laboratories created a system specifically for solving his bird puzzles, and it generated quite a lot of interest world-wide. Another work along these lines is *Forever Undecided* (Smullyan 1987), which discusses Gödel's theorem in greater detail that before, and also modal logic and provability logics. Raymond actually presents the underlying ideas of provability logic through puzzles in a way that makes them accessible to anybody from high-school up. More recently his book, *Satan, Cantor, and Infinity* (Smullyan 1992b) was republished by Dover, 2009. In this he discusses issues of probability, infinity, time, and change, largely through his familiar medium of puzzles. This is far from a complete list of Raymond's puzzle books.

Chess puzzles also intrigue Raymond, but only certain kinds of chess puzzles. Generally chess problems take the form illustrated by: white to move and mate in three, what are the moves. A Smullyan chess puzzle is *retrograde*. Instead of deducing where you are going, you must deduce where you have been. Here is a relatively simple example from (Smullyan 1979). Positions and colors are shown, except that for one pawn the color is not known. You are told that so far, no piece or pawn has ever moved from a square of one color to a square of the other. You are told nothing else, not even which side is to move. The problem is to determine the color of the ambiguous pawn. I repeat, this is a relatively simple example.

Raymond Smullyan
The Chess Mysteries of Sherlock Holmes
1979, pp 21–23

Raymond incorporated his puzzles into witty and entertaining stories, and these have been published as *The Chess Mysteries of Sherlock Holmes*, mentioned above, and *The Chess Mysteries of the Arabian Knights* (Smullyan 1982b). Considerable work, not by Raymond, has gone into developing computer programs capable of solving retrograde analysis puzzles, and Raymond's books have been a primary influence. Again, however, self-reference is not relevant for the solution of such puzzles, and we will not discuss them further here.

Raymond has belonged to both mathematics and philosophy departments over the years. Some of his philosophical work—rather non-traditional—has been collected in *5000 B.C. and Other Philosophical Fantasies* (Smullyan 1983), and *The Tao is*

Silent (Smullyan 1977). Again, these do not consider self-reference, and are merely mentioned in passing.

Raymond is in his nineth decade. This has not meant a lessening in productivity. In 2009 he published *Logical Labyrinths*, A. K. Peters. He refers to this as his 'bridge' book, because in it he bridges his puzzle world and his more traditional mathematical logic world, starting with puzzles and concluding with a thorough presentation of the basics of mathematical logic. In addition, in a more personal direction, there are *A Spiritual Journey* (Smullyan 2009a), *Rambles Through My Library* (Smullyan 2009b), and *Reflections* (Smullyan 2015).

As we have seen, Raymond's logical research and writing has a somewhat peculiar trajectory to it, beginning with important technical results in classical recursion theory, moving to influential books with a pedagogical purpose, then constantly circling back through both technical and popular output. The topic of self-reference is a common thread connecting much, but hardly all, of this work.

1.2 Self-Reference, the Mathematical Side

Raymond's writing on self-reference falls into two overlapping categories. Some of it is purely technical, some is meant for a more general audience. The same themes wander through both groups. In this section is a discussion of a small, but hopefully representative, selection from his more technical work. As to the popular writing, that is discussed in the next section, by an expert.

Of course self-reference occurs when something refers to itself. What kind of thing the something may be, and what it means to refer, are where the God of mathematics sprinkles the details. Sometimes these details can be complex, as with Gödel numbering, which assigns numbers to syntactical objects so that formal arithmetic can refer to its own expressions. Raymond's very first published paper was *Languages in which self-reference is possible*, (Smullyan 1957). This rather obviously deals with self-reference, but the basic motivation was to strip off as much of the formal machinery as possible, to get at the essence of what was going on. To give an idea of how far he took things with minimal machinery, here is a summary of part of the paper, near the beginning. The essential ideas are that some version of quotation is needed—one can think of Gödel numbering as a quotation mechanism. And one needs application, one expression applied to another. Loosely, this his how one expression 'talks about' another. In his paper application is written, simply, as one expression followed by another.

We begin with some simple syntax, embodied in a very elementary system called S_0. There are just three symbols: φ, *, and N. φ is a kind of predicate symbol which will be assigned a meaning when semantics is introduced. * acts like a quotation mark. And N stands for what Raymond calls a *norm* function, where a norm is roughly the application of an expression of S_0 to itself. The formal syntactical details for S_0 are these.

Syn-1 An *expression* of S_0 is any string built from φ, $*$, and N.
Syn-2 If E is an expression, $*E*$ is its *quotation*.
Syn-3 If E is an expression, $E*E*$ is its *norm*, that is, E followed by its own quotation.

Next, *designators* and *designation* are defined. The idea is that designators are expressions that designate other expressions (possibly themselves).

Des-1 If E is an expression, $*E*$ is a designator, and it designates E.
Des-2 If E_1 is a designator that designates E_2, then NE_1 is a designator, and it designates the norm of E_2, that is, it designates E_2*E_2*.

We have not given any role to φ yet. Syntactically it is treated as a predicate that can apply to designators. Formally, a *sentence* of S_0 is an expression of the form φE where E is a designator. Now a simple semantics is introduced, in which φ is explicitly interpreted as a property of designators.[4] This is embodied in the following definition of a *semantical system* S_P, where P is an arbitrary set of expressions. The idea is that P provides the meaning for φ.

Sem-1 If E is a designator, φE is true in S_P if the expression designated by E is in P.
Sem-2 if φE is not true in S_P, we say it is false in S_P.

That's it. And here are a few basic theorems. The first is purely syntactical; S_P does not come into it. But it shows that even with this minimal machinery we have self-reference.

Theorem A *There is an expression of S_0 that designates itself.*

Proof By Des-1 $*N*$ is a designator that designates N. Then by Des-2, $N*N*$ is a designator that designates the norm of N, but by Syn-3, that norm is $N*N*$. Thus $N*N*$ is a designator that designates itself. ■

Next the semantics is brought in. The following proof starts much like the previous one, with one more important step.

Theorem B *There is a sentence G of S_0 such that, no matter what the property P is, G is true in S_P if and only if G has property P (that is, $G \in P$).*

Proof By Des-1, $*\varphi N*$ is a designator, and it designates φN. Then by Des-2, $N*\varphi N*$ designates $\varphi N*\varphi N*$. Take G to be $\varphi N*\varphi N*$. Then G is true in S_P if and only if $\varphi N*\varphi N*$ is true in S_P and by Sem-1, this means the expression designated by $N*\varphi N*$ is in P, that is just when $\varphi N*\varphi N*$ is in P. In other words, just when $G \in P$. ■

---

[4] We follow the standard mathematical practice of identifying properties with sets, and application of a property as membership in the set. Still, we continue to use the word *property*.

Now there are two quick consequences that quickly get us to the heart of Gödel's and Tarski's fundamental theorems. There is a small subtlety that should be mentioned first, to account for the two versions. An expression φE that is false in S_P might be so for either of two reasons. It is false if E is a designator, but E is not in P. It is also false if E is not a designator, since only expressions of the form φE *where E is a designator* are considered true.

Corrollary C *For every P it is never the case that P is the set of all false sentences of S_P, and it is never the case that P is the set of all false expressions of S_P.*

Proof By Theorem B there is a sentence G such that G is true in S_P if and only if $G \in P$, whatever P may be. By Sem-2, this can be restated as: G is not false if and only if $G \in P$. This is impossible if P is either the set of false sentences or false expressions. ■

I stop this discussion here. In his paper Raymond went much further, to the point of applying his ideas to formal arithmetic. That would be beyond the scope of this introduction. I have merely given the flavor. Raymond's initial publication has much of technical interest, but probably its main contribution was the startling look it provided at the bare bones of the results of Gödel, Tarski, Church, and others. Sometimes God is *not* in the details.

Incidentally, Raymond has never stopped thinking about simplification. A more recent version of the ideas just discussed can be found later in this Introduction in *Self-Reference, the Popular Side*, the section immediately following this one.

Before considering another important early result of Raymond, a few words about self-reference and fixpoint theorems is in order. There are well-known connections, but well-known does not imply that it is known to everybody. Two examples are presented, one baby, the other more adult.

Russell's paradox in naive set theory is (almost) common knowledge. Let N be the set of all sets that do not belong to themselves. Symbolically, for any set x

$$x \in N \Leftrightarrow x \notin x$$

Then what is the status of the set N itself? Taking x to be N we get

$$N \in N \Leftrightarrow N \notin N$$

and this is a contradiction.

Self-reference is very obvious in this example. We have applied the definition of N to N itself. This application seems unproblemetic. We have used the definition in the way definitions of sets are always used, so why should asking about the set itself be a forbidden question? Still, we have a contradiction. It is generally understood that this contradiction rules out naive comprehension, that is, we cannot simply assume every condition determines a set, so in particular it cannot be that there is a set N.

In axiomatic set theory, naive comprehension is replaced with a more restricted version called separation: given a set a, each condition determines a *subset* of a. Using separation, for each set a we are allowed to form the set of all members of a that do not belong to themselves. Then we can meaningfully talk about the following function, defined on all sets.

$$f(a) = \{x \in a \mid x \notin x\}$$

That is, $f(a)$ consists of those members of a that are not members of themselves.[5]

We can now define the following function on all sets.

$$g(a) = a \cup \{f(a)\}$$

That is, $g(a)$ enlarges a by throwing $f(a)$ in as an additional member. Note that, for every a, $a \subseteq g(a)$, since a is explicitly part of the output. A set a such that $a \subseteq g(a)$ is sometimes called a *semi-fixed point* of g. Then *every* set is a semi-fixed point for g. Nonetheless, we can easily show that g has no fixed points, that is, it never happens that $a = g(a)$.

First we show a preliminary result: it never happens that $a \in f(a)$. We do this by showing that if $a \in f(a)$ we would have a contradiction. Here is the argument for this. From the definition of f,

$$x \in f(a) \Leftrightarrow x \in a \wedge x \notin x.$$

Then, taking x to be $f(a)$,

$$f(a) \in f(a) \Longleftrightarrow f(a) \in a \wedge f(a) \notin f(a).$$

Now, suppose it happened that $f(a) \in a$. Then the conjunction on the right of the double arrow could be simplified by dropping a conjunct we know to be true, and we would have

$$f(a) \in f(a) \Longleftrightarrow f(a) \notin f(a).$$

But exactly as with Russell's paradox, this is a contradiction. So it must be that we never have $f(a) \in a$.

But then g can not have a fixed point by the following simple argument. We must have that $f(a)$ is a member of $g(a)$ since $g(a)$ is $a \cup \{f(a)\}$, and the output explicitly includes $f(a)$. Since we have $f(a) \in g(a)$, if we had $a = g(a)$ we would have $f(a) \in a$ but as we have just seen, we cannot have this.

---

[5]Technically this function is not itself a set, but is characterized by a first-order formula in the language of set theory. We don't need such details for what we are doing now. It suffices to note that everything thus far is legitimate in generally accepted formal set theories.

Once again we can draw a conclusion about set theory from our paradoxical argument: there is no universal set U, that is, there is no set U of all sets. For, suppose there were a universal set U. We have seen that, for every set a, $a \subseteq g(a)$ so as a special case, $U \subseteq g(U)$. But since U is the universal set every set is a member of it. In particular, all the members of $g(U)$ must be members of U and so $g(U) \subseteq U$. It follows that $g(U) = U$, that is, U is a *fixed point* of g, and we know there is no such thing.

This example illustrates the interchangeability of notions of self-reference and fixed points, at least in certain contexts. This is a point that may be of use when reading some of the contributions to this volume.

Here is a second, more subtle example of a connection between self-reference and fixed points. As is well-known, Gödel showed the incompleteness of Peano arithmetic by making use of a formula that, informally understood, expresses its own unprovability.

Of course formulas are not what Peano arithmetic talks about, so what is now called *Gödel numbering* was introduced, assigning numbers to strings of symbols so that the usual syntactic operations like concatenation correspond to computable functions on their associated numbers. But Peano arithmetic does not even talk about numbers directly. It talks about *numerals*, strings of symbols that name numbers, as with our familiar place value number names. We can suppress all this detail here, and use the following notation. For a formula X, X° is the numeral designating the Gödel number of X.

Now, what Gödel constructed was a formula $\mathsf{Bew}(x)$ with one free variable that, in a plausible sense, embodies the idea that x is the numeral of the Gödel number of a proof in Peano arithmetic. He then showed there is a formula F with no free variables such that the following is provable in Peano arithmetic.

$$F \equiv \neg\mathsf{Bew}(F^\circ)$$

Intuitively, F 'says' that F is not provable. It is a famous example of self-reference. Carnap later observed that Gödel's construction actually was more general than he stated it. For any formula $\varphi(x)$ with one free variable, there must be a formula F with no free variables so that the following is provable in Peano arithmetic.

$$F \equiv \varphi(F^\circ)$$

This is a fixed point result. Gödel's instance is the case where $\varphi(x)$ is $\neg\mathsf{Bew}(x)$, but other choices have yielded many important results over the years.

The Gödel-Carnap fixed point theorem is in the language of formal arithmetic and concerns provability in the sense of mathematical logic. There is an analogous result in computability theory, the Kleene second fixed point theorem. It has two basic forms. It is often stated for computable functions, roughly those functions on non-negative integers (hereafter just called integers) for which an input/output computer program can be specified. It can also be stated for computably generated relations, again roughly, those sets and relations on integers whose instances can be generated by a computer program.[6] We have chosen to discuss the relation version, and not in its most general form at that.

Each computably generated relation is generated using some computer program. Such a program can be coded by an integer, in a way a kind of Gödel numbering of programs. Let ω_n be the computably generated set generated by the program coded by the integer n, called an *index* for ω_n. Suppose $R(x, y)$ is some computably generated binary relation. It can be shown that for each integer k the set $\{x \mid R(x, k)\}$ is a computably generated set, and hence has an index. (And similarly for relations that are more than binary.) Then for each k there is some n such that

$$\omega_n = \{x \mid R(x, k)\}.$$

The Kleene second recursion theorem (also discovered independently by Martin Davis) tells us that this relationship has a fixed point. That is, for each computably generated relation $R(x, y)$ there is an integer n such that $\omega_n = \{x \mid R(x, n)\}$. From this many remarkable results follow with astonishing ease. Indeed, with only a little more work a stronger version can be shown: such an n can be computed from an index for R, but we will not state this more formally.

The Gödel-Carnap arithmetic fixed point theorem and the Kleene second recursion theorem are, in a sense, different versions of a similar underlying phenomenon, and often it is possible to employ either for a particular purpose. For example, Solovay's original proof of the arithmetic completeness theorem for provability logic, *Provability interpretation of modal logic*, Solovay (1976) (which we won't discuss here) made a rather intricate use of the second recursion theorem, while in his book *The Unprovability of Consistency*, Boolos (1979), Boolos used the Gödel-Carnap theorem in a similar way.

With this background discussion out of the way, we move to one more example of Raymond's early work, a *Double* Recursion Theorem, originating in (Smullyan and Putnam 1960).

Suppose $R_1(x, y, z)$ and $R_2(x, y, z)$ are computably generated three-place relations. Then there are integers a and b such that we have both of the following.

$$\omega_a = \{x \mid R_1(x, a, b)\} \text{ and } \omega_b = \{x \mid R_2(x, a, b)\}$$

Again there is a stronger version that tells us a and b can be computed from indexes for R_1 and R_2.

Raymond's first application of this theorem was in proving that, up to computable isomorphism, there is only one pair of effectively inseparable sets. This sounds rather technical and I won't go into the details here. But the double recursion theorem itself is elegant, it's proof is ingenious, and it has had a long life. Its applications are many, including recent work on formalizing and studying the reproduction of self-replicating computer viruses.

[6]Older usage calls computable functions *recursive* functions, and computably generated relations *recursively enumerable* relations. In recent years terminology has shifted.

Here is an anecdote from long, long ago, saying something about how our minds work. Raymond once told me he was in the library at Princeton, a graduate student, and had been thinking for days about a possible double recursion theorem proof. He could feel that he was about to get it, but some students were making noise nearby. He asked them if they would be quieter because he was on the verge of getting an important idea. As I recall, he said they were, and he did.

Raymond moves from topic to topic, but nothing ever gets left behind. He revisited the double recursion theorem many times, each time making it part of a larger context. It appears in his books *Theory of Formal Systems*, (Smullyan 1961), *Recursion Theory for Metamathematics*, (Smullyan 1993), and *Diagonalization and Self-Reference*, (Smullyan 1994). Most recently, I interviewed him on April 1 of 2016, 56 years after the Double Recursion Theorem first appeared. I asked him about its origins and, without any hesitation on his part, received the following answer.

> Myhill proved, up to recursive isomorphism, there is only one creative set. I thought maybe I could prove, up to recursive isomorphism, there is only one pair of effectively inseparable sets. For that (Myhill's result) I imagine it is vital to use the recursion theorem. (So) I would need a double recursion theorem. I was motivated by that.
>
> I also used it (the double recursion theorem) to prove that given any disjoint pair of recursively enumerable sets there exists a formula provable for everything in the first set, refutable for everything in the second, and undecidable everywhere else.
>
> (What was Hillary Putnam's role in this?)
>
> I made a slight error, an oversight, which he corrected and I made it a joint paper.

Still in mind, still waiting for further development, for deeper understanding.

1.3 Self-Reference, the Popular Side

Raymond's wide influence, not surprisingly, has been more through his popular writings than through his technical results. His puzzle books have had an impact that continues and grows. I asked him if there was something particular he wanted me to say about this body of work. He replied, "Why don't I write something myself?" It is hard to resist an offer for self-reference from an expert. So, here is what he wrote.

My Involvement with Self-Reference
Raymond Smullyan

When I was six years old, I was in bed with a cold, or something, on April 1. My brother Emile, ten years older than I, came into my room in the morning and said: "Today is April Fools Day, and I will fool you like you have never been fooled before!" Well, I waited all day long for him to fool me, but he didn't. When night came, my mother asked me why I didn't go to sleep. I replied that I was waiting for Emile to fool me. My mother said to Emile, "Why don't you fool the child?" Emile then turned to me, and the following dialogue ensued.

Emile: You expected me to fool you, didn't you?

Raymond: Yes.
Emile: But I didn't, did I?
Raymond: No.
Emile: But you expected me to, didn't you?
Raymond: Yes.
Emile: So I fooled you, didn't I?

I recall then lying in bed wondering whether I had been fooled our not. If I wasn't fooled, but expected to be, then was I not then fooled? This paradox really grabbed my attention.

A related incident: Many years later, I thought of earning some money as a salesman. I applied to a company and had to take an aptitude test. One of the questions was, "Do you object to telling a little lie every now and then?" Well, I did object, but thought that if I said so, I wouldn't get the job, hence I lied and said NO. But doesn't this cause a paradox? I didn't object to the lie I told, hence I don't object to sometimes lying, therefore my answer NO was not a lie, but the truth! And so, was I lying or not?

As one mathematician told me, these two incidents may well be forerunners in my later interest in Gödel's theorem and self-reference.

I like to illustrate a key idea behind Gödel's theorem as follows: Imagine a computing machine that prints out expressions composed of the following three symbols:

$$P, R, N$$

By a *sentence* is meant any expression of one of the following four forms (where X is any expression in these symbols whatsoever).

1. PX
2. NPX
3. RPX
4. NRPX

These sentences are interpreted as follows.

1. P stands for *printable* (by the machine) and so PX is called *true* iff (if and only if) X is printable.
2. N stands for *not*, and so NPX is called *true* iff X is not printable.
3. R stands for *repeat*, and so RPX is called *true* iff XX (the repeat of X) is printable.
4. Again, N stands for *not*, and so $NRPX$ is called *true* iff the machine cannot print XX.

We are given that the machine is wholly accurate, in the sense that every sentence printed by the machine is true. Thus if PX is printable, then so is X. Is the converse necessarily true? If X is printable, does it necessarily follow that PX is printable? No, it does not! If X is printable, then PX is certainly true, but that does not mean that PX is printable! I never said that all true sentences are printable, but only that all printable sentences are true. As a matter of fact, there is a true sentence that is not printable. Can the reader find one? [It bears a close relation to Gödel's famous undecidable sentence. A solution will be given later.]

Incidentally, this machine has another curious feature: it is possible to construct two sentences X and Y such that one of the two must be true but not printable, but there is no way of telling which one it is! [There are two solutions and they are given later. The idea is to construct them such that X asserts (i.e. is true iff) Y is printable, and Y asserts that X is not printable.]

I must now tell you of an amusing incident. In one of my puzzle books, I gave a proof that either Tweedledum exists, or Tweedledee exists, but there is no telling which. This, together with my double machine puzzle above, inspired Professor Melvin Fitting to once introduce me at a mathematics talk as follows. He said: "I now introduce Professor Smullyan, who will prove to you that either he doesn't exist, or you don't exist, but you won't know which!"

Now for the solutions of the two puzzles. We recall that for any expression X, the sentence $NRPX$ is true iff the repeat of X is not printable. Well, we take NRP for X, and so $NRPNRP$ is true iff the repeat of NRP is not printable, but the repeat of NRP is the very sentence $NRPNRP$! Thus $NRPNRP$ is either true and not printable, or false but printable. The later alternative is ruled out by the given condition that only true sentences are printable. Thus the sentence $NRPNRP$ is true, but the machine cannot print it.

Remarks

1. The fact that no sentence is printable which asserts that it cannot be printed is reminiscent of a scene in Shakespeare's "Romeo and Juliet," in which the nurse runs to Juliet and says "I have no breath," and Juliet responds by saying, "How can you say you have no breath, when you have breath left to say 'I have no breath'?"
2. The sentence $RPNRP$ is false (since $NRPNRP$ is true) hence is also unprintable. Thus neither $RPNRP$ nor its negation $NRPNRP$ are printable. If we think of the printable sentences as the *provable* sentences of some mathematical system, then the sentence $RPNRP$ is an *undecidable* sentence of the system, and resembles Gödel's famous undecidable sentence.

Now for the second problem. Suppose X and Y are sentences such that X is true iff Y is printable, and Y is true iff X is not printable. Well, X is either true or false. Suppose X is true. Then Y is printable, as X truly says. Hence Y must be true (by the assumption that only true sentences are printable). Hence X is not printable (as Y truly says). Thus X is true but not printable.

Now consider the case that X is not true. Then, contrary to what X says, Y is not printable. Yet Y is true, since it correctly asserts that X is not printable. And so in this case, Y is true but not printable.

In summary, if X is true, then X is true but not printable, but if X is false, then it is Y that is true but not printable. There is no way to tell which of these two cases hold.

Now, how do we get two such sentences X and Y? One solution is as follows.

$$X = PNRPPNRP$$
$$Y = NRPPNRP$$

Since $X = PY$, then obviously X is true iff Y is printable. Also Y is true iff the repeat of $PNRP$ is not printable, but the repeat of $PNRP$ is X.

Another solution is the following.

$$X = RPNPRP$$
$$Y = NPPRNPRP$$

X is true iff the repeat of $NPRP$ is printable, but this repeat is Y. Also Y, which is NPX, is true iff X is not printable.

Gödel's incompleteness proof combined a self-referential device with an issue about ω-consistency, and I believe these two things could and should be separated out, which I will now do.

Let $\varphi(x)$ be a formula with just the one free variable x, and for any (natural) number n, $\varphi(\overline{n})$ is the result of substituting the name $\overline{n}$ of n for all free occurrences of x in $\varphi(x)$. We say that $\varphi(x)$ *represents* the set of all (natural) numbers n such that $\varphi(\overline{n})$ is provable (in the system under consideration).

If a number n is the Gödel number of a formula $\varphi(x)$ I denote such a formula as $\varphi_n(x)$.

As with my other writings, I let P^* be the set of all n such that $\varphi_n(\overline{n})$ is provable. Gödel's incompleteness proof for such systems as Peano Arithmetic boils down to representing the set P^*. This guarantees incompleteness (assuming simple consistency) as follows: suppose $\varphi_n(x)$ represents P^*. Let k be the Gödel number of the formula $\neg\varphi_n(x)$—thus $\varphi_k(x) = \neg\varphi_n(x)$. Then $\varphi_n(\overline{k})$ must be undecidable (assuming the system is simply consistent) because: $\varphi_n(\overline{k})$ is provable iff $k \in P^*$ iff $\varphi_k(\overline{k})$ is provable iff $\varphi_n(\overline{k})$ is refutable (i.e. its negation $\neg\varphi_n(\overline{k})$ is provable). Thus $\varphi_n(\overline{k})$ is provable iff it is refutable, and so if the system is simply consistent (no sentence is both provable and refutable), $\varphi_n(\overline{k})$ must be undecidable, neither provable nor refutable.

We note that the proof above nowhere required the hypothesis of ω-consistency, but Gödel required that hypothesis for his proof. And so, where does ω-consistency enter the picture? Well Gödel required the hypothesis of ω-consistency in order to show that P^* is indeed representable! Here is how.

Let $\psi(x, y)$ be a formula with just x and y as the free variables, and for any numbers m and n, $\psi(\overline{m}, \overline{n})$ is the result of substituting $\overline{m}$ for all free occurrences of x and $\overline{n}$ for all free occurrences of y in $\psi(x, y)$.

Given a relation $R(x, y)$ between natural numbers, the formula $\psi(x, y)$ is said to *define* (or *numeralwise express*) the relation R if, for all numbers m and n, the following two conditions are realized:

1. If $R(m, n)$ holds, then $\psi(\overline{m}, \overline{n})$ is provable.
2. If $R(m, n)$ doesn't hold, then $\psi(\overline{m}, \overline{n})$ is refutable (its negation is provable).

By the *domain* of R we mean the set of all m such that $R(m.n)$ holds for at least one number n. Now without the assumption of ω-consistency, Gödel managed to construct a formula $\psi(x, y)$ which defined a relation whose domain was P^*, but to pass from this to the representability of P^*, Gödel required the hypothesis of ω-consistency (which is that if a formula $\varphi(x)$ is such that $\exists x\varphi(x)$ is provable, then it is *not* the case that all the sentences $\varphi(\overline{0})$, $\varphi(\overline{1})$, $\varphi(\overline{1})$, ... $\varphi(\overline{n})$, ... be refutable.

More generally, let us note the following important fact: given a relation $R(x, y)$ with domain A, and a formula $\psi(x, y)$ that defines R, the formula $\exists x\psi(x, y)$ will represent A providing the system is ω-consistent! Here is why. Assume $\psi(x, y)$ defies R. If $m \in A$, then $R(m, n)$ holds for at least one number n, and therefore $\psi(\overline{m}, \overline{n})$ is provable, hence so is the sentence $\exists y\psi(\overline{m}, y)$. Thus the set represented by the formula $\exists y\psi(x, y)$ includes all the members of A, but it might include more. But if the system is ω-consistent, it cannot include more, for suppose $\exists y\psi(\overline{m}, y)$ were provable for some m outside A. Then $R(m, n)$ fails for every n, hence $\psi(\overline{m}, \overline{n})$ is refutable for every n, hence the system would be ω-inconsistent (since $\exists y\psi(\overline{m}, y)$ is also provable). Thus if the system is ω-consistent, then $\exists y\psi(x, y)$ does indeed represent the domain of R.

1.4 What Must Be Said

Dear Reader, you have just been reading Raymond Smullyan's own contributions to this very book honoring Raymond Smullyan. It is a truly apt example of self-reference, from the master. But there will be no more. Raymond passed away on February 6, 2017 at the age of 97, while this book was in production. I am happy to report that when the manuscript was sent to the publisher, a copy was given to Raymond and he had the opportunity to enjoy what his friends and colleagues said. He was in no position to judge this book by its cover, but he wouldn't have anyway. His judgment was simply one of pleasure and gratitude to those who contributed. And we contributors are pleased and grateful to have had the opportunity to contribute. We will miss him.

1.5 Bibliography

Details concerning works by Raymond Smullyan that are cited above can be found in the Smullyan Bibliography at the end of this book. References at the end of this chapter are the cited works that are by others.

1.6 Papers in This Volume

Papers are grouped loosely and somewhat informally into similarity type.

Purely Academic Papers

- Robert L. Constable: *Formal Systems, Logics, and Programs.* Proof assistants are, in a sense, software assistants for humans attempting to provide correct proofs, especially proofs having to do with software verification, though they are very useful for more 'mathematical' applications as well. In Robert Constable's article, two successful uses of the proof assistant Nuprl are discussed in an informal way that should appeal to an audience of non-specialists. The first is the formalization of a fundamental result from Raymond's book *First-Order Logic*: an assignment of truth values to propositional variables extends to at most one assignment of truth values to all propositional formulas that respects the logical connectives. What is significant here is that in his book Raymond gives this a very special constructive proof. This constructive existence proof, formalized using 'very dependent function types,' allows one to extract a recursive evaluation function. The second involvement of Nuprl has to do with the formalization of the halting problem in constructive type theory, by making use of the important concept of *partial types*. The key thing in both these is that the discipline imposed by the use of a proof assistant results in a better understanding of the constructive content of arguments.
- Robert Cowen: *Adaptive Fault Diagnosis using Self-Referential Reasoning.* Processors (components, chips, etc.) are much like Raymond's Knights, who are truth-telling, Knaves, who lie, and Normals, who sometimes lie and sometimes tell the truth. The aim of this paper is to make use of self-referential questions, Smullyan style, to determine the reliability of processors at a distance, say in space or in a nuclear reactor. For this purpose, both Knights and Knaves are considered reliable. If we know a processor is Knave-like, we know how to make use of its answers. It is the Normals who are the problem. Cowen provides an algorithm for excluding Normals using self-referential questions, subject to the condition that more than half the processors are reliable. This is a curious combination of logic, combinatorics, and an abstract version of a real-world problem.

- Melvin Fitting, *Russell's Paradox, Gödel's Theorem* This section of the introduction is supposed to consist of summaries of the various contributions to the present volume. (How's that for self-reference?) Since this is my own contribution, I think the abstract I wrote already in the paper accomplishes the purpose, so I reproduce it here.
  Smullyan has closely examined Gödel's proof of arithmetic incompleteness both abstractly, for a mathematically sophisticated audience, and via the medium of puzzles, for a general audience. In this essentially pedagogical paper I take an intermediate position, and show that Gödel's ideas fit naturally into the context of the elementary set theory that any mathematics student would know near the beginning of studies. Gödel once wrote, about his incompleteness proof, "The analogy of this argument with the Richard antinomy leaps to the eye. It is closely related to the 'Liar' too." And further, "Any epistemological antinomy could be used for a similar proof of the existence of undecidable propositions." Here, combining the spirit of both Smullyan's formal and his popular work, I show how Russell's paradox can be used in a direct and natural way to obtain Gödel's incompleteness results.

Puzzle Oriented Academic Papers

- Henk Barendregt, Jörg Endrullis, Jan Willem Klop, and Johannes Waldmann: *Dance of the Starlings*. Combinatory logic, and its close relative the lambda calculus, have to do with the seemingly innocuous business of substituting one expression for another in a complex context. Remarkably, this is a much more complicated issue than one might think, and has bearing on both computability theory and on the design of programming languages. In *To Mock a Mockingbird*, (Smullyan 1985), Raymond presented the ideas of combinatory logic through a series of puzzles about birds. The application operation of combinatory logic turned into one bird calling to another. There was a mocking bird, able to mimic other birds. And so on. In *Dance of the Starlings* a number of important features of the basic combinator S are examined. Whimsically, S could stand for the bird starling, or for Smullyan, or for Schönfinkel, who invented combinatory logic in the first place.
- Rick Statman: *Some Tweets About Mockingbirds*. Like *Dance of the Starlings*, this is another paper about combinatory logic, deriving from Raymond's *To Mock a Mockingbird*. In his book Raymond asked whether a fixed point combinator could be constructed from his Bluebird, Identity bird, and Mockingbird combinators B, I, and M. This is still not known, but the present paper is an examination of some closely related questions that are interesting in their own right.
- Stephen Yablo: *Knights, Knaves, Truth, Truthfulness, Grounding, Tethering, Aboutness, and Paradox*. Raymond's Knight/Knave puzzles rely heavily on self-referential statements. In this paper, which is of both philosophical and technical interest, recent theories of truth and self-reference are examined for what they have to say about an issue which is, as the paper says, a "problem not apparent even to veteran paradox-mongers" and "is a datum in need of explanation." The paper makes the problem properly apparent, and provides a proposal for a solution.

General Essays

- Martin Davis: *What I Tell You Three Times Is True*. This whimsical article makes a connection between Raymond's use of self-reference and self-referentiality in art. Here is an example to serve as an advertisement for the paper, but one that does not occur in it. The American painter Norman Rockwell, famous for his magazine covers, once painted a self-portrait. In it you see the artist from behind, sitting awkwardly while painting his own portrait. You also see his somewhat mousy reflection in a mirror. He is basing his painting on this. And finally you see the painting he is making, in which he looks younger, striking, and distinguished. Altogether, three versions of Norman Rockwell—perhaps relatives of Raymond's Knights, Knaves, and Normals?
- Vann McGee: *J. R. Lucas and the Soul-Searching Selfie.* In this paper, Mechanism is understood to be "the doctrine that the symbol-producing capabilities of the human mind can be simulated by a finite machine." This is, of course, weaker than simply asserting that the mind *is* a machine. But it is a doctrine that has gathered about it many attempts to say something rigorous in its defense, or the opposite. Familiar names here include Lucas, Penrose, and Putnam, all of whom proved something about Mechanism, or didn't, or maybe what was proved wasn't exactly about Mechanism after all. In this very readable paper, McGee examines the Mechanism issue again. It is a lucid, convincing paper, and will not settle the debate because it is unlikely that anything will.

Puzzles

- Andrew G. Buchanan and John H. Conway: *A Sea Island Tale for Young Anthropologists*. Perhaps the most accurate way to describe this contribution is that it is one more Raymond Smullyan Knight/Knave puzzle book, except that it's not a book, and it's not by Raymond. Told as a typical Smullyan story, filled with very ingenious puzzles, it concludes with a formidable list of puzzle complications that remain to be explored.
- Walter Carnielli: *Making the 'Hardest Logic Puzzle Ever' Endlessly Harder*. Raymond is famous for his logic puzzles, very commonly involving self-referentiality. In "The hardest logic puzzle ever" (*The Harvard Review of Philosophy* 6, pp. 62–65, 1996), George Boolos introduced what he claimed was the hardest logic puzzle ever, which he credited as originating with Raymond Smullyan (though it is not one that appears in his books) but with an important complication added by John McCarthy. There are three gods, one a liar, one a truth teller, and one who answers randomly. Of course you are not told which are which. In addition, answers to yes/no questions are given in a language that you don't understand. You must figure out who is the truth teller, who is the liar, and who is the randomizer, by asking three yes/no questions. Over the years the puzzle has generated an extensive literature. Because of the random answerer, three valued logic is implicit in the formulation of the puzzle, but solutions are usually considered to be based on classical logic. In *Making the 'Hardest Logic Puzzle Ever' Endlessly Harder*, Walter Carnielli considers the explicit use of three-valued logics not only for solving

the puzzle but for producing versions that, in some sense, are still harder. Since hardness is somewhat subjective, this is a claim about which the reader may form his or her own opinion, but the enterprise will certainly be sufficiently hard to puzzle to a reader's satisfaction.

1.7 Biographies of Authors and Editors

Authors are listed Alphabetically.

Henk Barendregt (1947) obtained his PhD (1971) in mathematical logic under supervision of Dirk van Dalen and Georg Kreisel at Utrecht University on the topics of lambda calculus and combinatory logic. From 1986 until 2015 he occupied the chair Foundations of mathematics and computer science at Radboud University. Besides work on lambda calculus, he has worked on automated certification of mathematics through formalization of proofs. In the present second phase of his scientific career Barendregt is focussing on consciousness, in particular on mind-states and their wholesome development towards insight based on concentration and mindfulness. The methodological approach here is phenomenological based on vipassana meditation. Barendregt is member of the Royal Dutch Academy of Science (1997), is recipient of the Spinoza Award (2002), was distinguished Lorentz Fellow at the Netherlands Institute of Advanced Studies (2012) and obtained a doctorate h.c. from Heriot Watt university (2015).

Andrew Buchanan is nomadic. British by origin he is currently resident in Hong Kong, where he is CEO of an innovative anti-counterfeiting start-up. He can vaguely remember studying mathematics in Cambridge UK, but he has worked in Information Technology for most of his career. Companies where he has fulfilled senior technical roles in mission critical programs include Bank of Ireland, Sappi, Winbond Electronics, Far EasTone, Hong Kong Jockey Club, Kraft, British Telecom, Royal Sun Alliance, and JP Morgan. As a schoolboy, he enjoyed Raymond Smullyan's *The Chess Mysteries of Sherlock Holmes*, and he has since published over 200 chess problems, mainly in retrograde analysis. His notorious discovery in 2000 of what became known as "Dead Reckoning" broke thousands of chess problems, finally resulting in a rules change by the World Federation of Chess Composition at their August 2015 Congress.

Walter Carnielli is professor of logic and philosophy at the Department of Philosophy, and also Director of the Centre for Logic, Epistemology and the History of Science of the University of Campinas, Brazil. He received a PhD in Mathematics from the University of Campinas and held research positions at the University of São Paulo, University of California Berkeley, University of Münster, University of Bonn, Superior Technical Institute Lisbon, and Université du Luxembourg. He has published several books and articles on combinatorics, proof theory and semantics for nonclassical logics, recursion theory and computability, set theory, modal logics, combinations of logics, and foundations of paraconsistent logics.

His concerns involve the expansion of reasoning horizons by non-classical logics, including the connections between logic, probability and game theory, as well as the philosophical interpretation of non-classical logics. He is executive editor of the *IGPL Journa*l (Oxford Journals), editor of "The Non-classical Logics Corner" of the *Journal of Logic and Computation* (together with Heinrich Wansing) and member of the editorial board of several other journals. He is a recipient of the Jabuti Prize, Brazil's most prestigious literary prize.

Robert Constable was an undergraduate in mathematics at Princeton University where Prof. Church supervised his senior thesis on set theory and Hilbert's epsilon operator. He did his PhD with Prof. Kleene in mathematical logic and a minor in computer science under Prof. Travis. He joined the Cornell Department of Computer Science in 1968 where he began research on computational complexity theory with his colleagues J. Hartmanis and J. Hopcroft.
He was drawn into designing and implementing programming logics in response to the "software crisis" of the 1970s. This work led to the idea of treating constructive proofs as programs which then led to assembling a group of PhD students to design and implement the first constructive type theory as a foundational theory for computer science. It became the logic of the Nuprl proof assistant.
During a sabbatical at the University of Edinburgh he worked with Robin Milner and learned the LCF system. He and PhD student Joseph Bates integrated the LCF tactic mechanism into Nuprl, which became operational in 1984 and is still used in implementing elements of mathematics and creating correct by construction programs and protocols. He has so far supervised forty three PhD students in computer science including Ed Clarke who won the Turing Award.

John H. Conway is the John von Neumann Distinguished Professor Emeritus of Applied and Computational Mathematics at Princeton. He has four doctorate degrees: PhD (Cambridge), DSc (Liverpool), D.h.c. (Iasi), PhD (Bremen). He has written or co-written 11 books, one of which has appeared in 11 languages. He has both daughters and sons, each daughter has as many sisters as brothers and each son twice as many sisters as brothers. He has met Raymond Smullyan repeatedly at many Gatherings for Gardner and Andrew Buchanan in Cambridge, New York and Princeton.

Robert Cowen is Professor Emeritus at Queens College, CUNY where he taught Mathematics for over 40 years. His research interests are Logic, Set Theory, Combinatorics, Theoretical Computer Science and Mathematica Programming. He first met Raymond Smullyan as an undergraduate at Princeton when he was a student in Ray's Logic course. He eventually followed Ray to Yeshiva University where Ray became his Ph.D. thesis advisor.

Martin Davis, born in New York City in 1928, was a student of Emil L. Post at City College and his doctorate at Princeton in 1950 was under the supervision of Alonzo Church. Davis's book *Computability and Unsolvability* (1958) has been called one of the few real classics in computer science. He is best known for his pioneering work in automated deduction and for his contributions to the solution of Hilbert's tenth problem for which latter he was awarded the Chauvenet and Lester R. Ford Prizes by the Mathematical Association of America and the Leroy

P. Steele Prize by the American Mathematical Society. In 1983 he was a Guggenheim Foundation Fellow. He received the Herbrand Prize from the Conference on Automated Deduction in 2005. His books have been translated into a number of languages including Russian and Japanese. Davis has been on the faculty of the Courant Institute of Mathematical Sciences of New York University since 1965, was one of the charter members of the Computer Science Department founded in 1969, and is now Professor Emeritus. He is a Visiting Scholar at the University of California, Berkeley.

Jörg Endrullis (1981) studied computer science at Leipzig University where his master's thesis has been supervised by Johannes Waldmann. Since 2005 he is connected to the Vrije Universiteit Amsterdam where he obtained his Ph.D. in 2010 with a thesis *Termination and Productivity* supervised by Jan Willem Klop and Roel de Vrijer. In 2013 he has worked 6 months at Indiana University Bloomington in the group of Larry Moss. He is currently working on term rewriting and lambda calculus, and infinite streams.

Melvin Fitting was born in Troy, New York, in 1942. His undergraduate degree was from Rensselaer Polytechnic Institute, in mathematics, and his 1968 PhD was supervised by Raymond Smullyan, at Yeshiva University. His dissertation became his first book, *Intuitionistic Logic, Model Theory, and Forcing* (1969). Much of his work has involved developing tableau systems for non-classical logics, thus generalizing the classical systems of his mentor Smullyan. In 2012 he received the Herbrand Award from the Conference on Automated Deduction, largely for this work. He was on the faculty of the City University of New York from 1969 to his retirement in 2013, at Lehman College, and at the Graduate Center, where he was in the Departments of Mathematics, Computer Science, and Philosophy. He has authored or co-authored 9 books as well as numerous research papers, covering philosophical logic, computability, automated theorem proving and, with Raymond Smullyan, set theory. He is currently an emeritus Professor, and very much active.

Jan Willem Klop (1945) studied mathematical logic at Utrecht University, obtaining his Ph.D. in 1980 with the thesis *Combinatory Reduction Systems* supervised by Dirk van Dalen and Henk Barendregt. Since 1980 he is connected to CWI Amsterdam and since 1986 Professor of Applied Logic at VU University Amsterdam. From 1998 until 2007 he was part-time connected to the group Foundations, led by Henk Barendregt at Radboud University Nijmegen. From 2002 to 2004 he was head of the department of computer science of the Vrije Universiteit in Amsterdam. In 2002 he received an honorary doctorate at the University of East Anglia, England, with Ronan Sleep as promotor h.c. Since 2003 he is a member of the Royal Netherlands Academy of Arts and Sciences (Koninklijke Nederlandse Akademie van Wetenschappen, KNAW), section Mathematics. Since 2005 he is a fellow of CWI Amsterdam. In 2011 he was appointed to honorary member of IFIP Working Group 1.6. Term Rewriting. In 2012 he was appointed to member of the Academia Europaea, section Informatics. Since 2011 he is emeritus of the Vrije universiteit Amsterdam, writing a book on term rewriting and lambda calculus, and one on infinite streams, with coauthor Jörg Endrullis.

Vann McGee was born in North Carolina during the postwar baby boom, studied at Harvard, the University of North Carolina, and Berkeley and taught at the University of Arizona and Rutgers before joining the Department of Linguistics and Philosophy at MIT. He works in philosophical logic, philosophy of language (vagueness especially), and philosophy of math.

Brian Rayman works for New York City's Department of Education. He holds Mathematics Education and Public Policy degrees from New York University and Columbia University, respectively. Since his days as an undergraduate, Brian has deeply embraced Professor Smullyan's works in recreational mathematics and Eastern philosophy. As a high school mathematics teacher, he integrated Raymond's work into his curriculum. In 2008, he invited Raymond to speak to his senior mathematics students. Raymond kindly accepted, and after two hours Brian's students were no longer sure where mathematics ends and magic begins, and vice versa. Today, he considers significant accomplishments as being counted amongst Raymond's friends and performing magic tricks that nearly make Raymond's eyeballs fall out. He lives in Carroll Gardens, Brooklyn, with his wife and two sons.

Rick Statman takes a minimalist approach to biography requests.

Johannes Waldmann (1967) studied mathematics at the University of Jena, Germany. In his Ph.D. thesis (1998), he showed that it is decidable whether a term built from S combinators is normalizing - solving a question asked by Jan Willem Klop decades ago. He then held a position at University of Leipzig; since 2003 he is a professor at University of Applied Sciences Leipzig. His research area is automated analysis of termination and derivational complexity of rewriting systems, using ideas from weighted automata, and implementing them with Constraint Programming. With co-author Dieter Hofbauer, their paper "Termination of String Rewriting with Matrix Interpretations" got the best paper award at the International Conference on Rewriting Techniques and Applications (RTA) 2006. He takes part in Termination Competitions since 2003, is a member of this competitions's steering committee, and this competition's organizer since 2014. He was also the organizer of the Programming Contest associated with the International Conference on Functional Programming in 2010, and used this to crowd-source the generation of hundreds of new hard termination problems. Together with Alfons Geser, he hosted the International Workshop on Termination in 2009, and the International School on Rewriting in 2015.

Stephen Yablo is David W. Skinner Professor of Linguistics and Philosophy at the Massachusetts Institute of Technology. He works in metaphysics and philosophy of mind and language. He is the author of *Thoughts: Papers on Mind, Meaning, and Modality* (OUP, 2008), *Things: Papers on Objects, Events, and Properties* (OUP, 2010), and *Aboutness* (Princeton University Press, 2014).

References

Boolos, G. (1979). *The Unprovability of Consistency*. Cambridge University Press.
Solovay, R. M. (1976). Provability interpretation of modal logic. *Israel Journal of Mathematics*, *25*, 287–304.

Chapter 2
Formal Systems, Logics, and Programs

Robert L. Constable

Abstract This article connects certain writings of Raymond Smullyan on logic and formal systems with current research on constructive type theory. It also considers aspects of teaching logic to computer science students. Those students require logic in order to precisely define programming tasks and demonstrate that programs meet logical specifications of these tasks. Smullyan's book *First-Order Logic* has been used for many years to teach logic to computer science students at Cornell. This article includes a brief account of an elegant result from this book. It is based on an extension of a fourteen page technical report that analyzes two pages on Boolean valuations in chapter one of his classic book. The analysis and results are formulated in a formal system of type theory and proved using the Nuprl proof assistant. The article also briefly considers the role of *partial types*, showing how they provide a new way to prove unsolvability results, another topic on which Smullyan has written with remarkable clarity. Partial types also provide a simple and faithful semantics for Hoare's *partial correctness* logic, widely used to reason about programs and taught in computer science as an important applied logic. They also make it possible to state a clean rule for defined functions in Hoare logic.

2.1 Introduction

Logic, Mathematics, and Computer Science Logic is perhaps the oldest established academic discipline. It plays an important role in shaping one of the newest disciplines, computer science. To oversimplify yet make the point, we note that *formal systems* came from Hilbert, *types* from Russell and Church, *algorithms* from Church and Turing, and *computation theory* from Kleene. These are core elements of modern computer science. This dynamic between logic and computer science

R. L. Constable (✉)
Cornell University, Ithaca, USA
e-mail: rc@cs.cornell.edu

M. Fitting and B. Rayman (eds.), *Raymond Smullyan on Self Reference*,
Outstanding Contributions to Logic 14, https://doi.org/10.1007/978-3-319-68732-2_2

persists, especially in the development and application of proof assistants for type theory, as we will illustrate in this article. First we note an interesting feature of the history of computer science in America.

Two major incubators for American computer science were located in Princeton – the Institute for Advanced Study and the Mathematics Department of Princeton University. Gathered there were Church, Turing, Kleene, Rosser, Gödel, von Neumann, McCarthy, and others. Much later, Raymond Smullyan was a Princeton mathematics graduate student working with Alonzo Church. I was an undergraduate mathematics major, and Professor Church supervised my senior thesis. At Princeton I learned of Smullyan's work in logic and heard stories about him being a magician.[1] I remain fascinated by Smullyan's work until this day, it has a "magical" quality. I have used his elegant book, *First-Order Logic* Smullyan (1968) in classes at Cornell for over twenty years – helping persuade Dover to continue publishing after it went out of print with Springer because I needed it for teaching. We also use *Logic for Applications* Nerode and Shore (1994) by Anil Nerode and Richard Shore which draws on *First-Order Logic* and other books by Smullyan.

Many have been inspired by Smullyan's work on self-reference Smullyan (1987), Smullyan (1994) and by his approach to Gödel's theorems Smullyan (1992). His popular works on this subject are very instructive as well Smullyan (1983), Smullyan (1984). His Princeton monograph on the theory of formal systems Smullyan (1961) was one of the first technical works I read that was not a textbook. It was readily available in the university book store. The account of the halting problem given here is in the spirit of Smullyan's work on self-reference.

From the book *First-Order Logic* I also became fascinated with tableau logics and learned the elegant formalization of intuitionistic logic from Fitting's book *Intuitionistic model theory and forcing* Fitting (1969). The logic of the Nuprl proof system Constable et al. (1986) is called a *refinement logic*. It is essentially a tableaux style logic as illustrated in the work of Underwood Underwood (1993). This article will also discuss a connection between the constructive type theory (CTT) of Nuprl and issues with the well known Hoare logic of programs O'Donnell (1982), Hoare (1971). The notion of partial types from CTT explains Hoare's rules at a denotational level, more abstract than the normal step by step account of computation used to prove rules correct in the completely formalized account of Hoare logic in the on-line book *Software Foundations* Pierce et al. (2013). Great care must be taken to avoid the well known inconsistencies in the rules for functions and procedures in Hoare logic O'Donnell (1982). We look at one example in Sect. 2.3.1. On the other hand, to teach these rules and make them intuitive, it is important to find the right abstractions. That is a lesson so well taught in all of Smullyan's work. It is a lesson that is increasingly important as hundreds of people use proof assistants to formalize elements of logic, mathematics, and computer science.

Focus of the article As stated in the abstract, this brief article discusses two topics, one directly related to the book *First-Order Logic*. That topic is treated by reporting on

[1] I had also assisted my uncle who was a magician, so I was fascinated by this connection to Smullyan already as an undergraduate.

a successful effort to formalize in Nuprl a simple but key theorem from the book. The theorem has implicit computational meaning that becomes explicit in an implemented formal constructive proof. The computational content of the proof can be executed by Nuprl. The account here is a summary of an unpublished technical report Allen et al. (2004) and results from a longer article, Polymorphic Logic Bickford and Constable (2012). In the short published article, my colleague Mark Bickford and I discuss the formalization of Smullyan's theorem on Boolean evaluation from *First-Order Logic* by using a type available in CTT, the *intersection type*. We formalized this result because it is elegant, and because it was a considerable challenge to make the formalization closely match the proof in the book – it took two tries Allen et al. (2004), Bickford and Constable (2012).

The second topic is inspired by Smullyan's interest in reducing results to their clearest formulations. I believe that the way we treat the *halting problem* in constructive type theory is particularly clear and simple. It requires the idea of a *partial type*. Early on Constructive Type Theory Constable et al. (1986) introduced partial types. They provide an appealing way to deal with problems in the formalization of Hoare Logics for programming and to formalize a constructive account of Dana Scott's domain theory Scott (1976), Crary (1998). Constructive domain theory was used by Karl Crary to provide the semantics for a realistic programming language, KML Crary (1998). This article suggests a new way of repairing false rules proposed for Hoare logics without resorting to the mechanism of "step by step semantics" that is more low level and costly. We cite the analysis by O'Donnell O'Donnell (1982) of one false rule proposed for functions in Hoare logic and show how the notion of a partial type repairs the error.

Background on type theory *Proof assistants* are used by many research and development projects worldwide to build verified software. Their use has been steadily growing over the past thirty years. These increasingly effective assistants integrate specification languages, programming languages, and programming logics to seamlessly support the coding and the verification tasks central to computer science and software engineering. This integration is especially useful when the specification language is rich and includes types natural to both *programming and logic*. Rich type systems allow precise mathematical specifications for a wide variety of computational problems. Moreover, as is now well known, *dependent types* not only express logic, they also generalize important programming types, e.g. from function types to *dependent function types*, from records to dependent records, from product types to subset types (refinement types), from polymorphism to uniformity, and so forth.

Proof assistants for type theory and higher order logics such as Agda Bove et al. (2009), Coq Bertot and Castéran (2004), HOL Gordon and Melham (1993), Nuprl Allen et al. (2006), MetaPRL Hickey et al. (2003), and Minlog Benl et al. (1998) have been in use for many years – for Nuprl thirty one years, the current version of Coq twenty years, HOL for over twenty years. For several reasons, the use of proof assistants has increased significantly over the past decade, to involve many hundreds of expert users. One reason is their record in building *provably correct* hardware and software: from compilers, OS kernels, distributed protocols, software defined networks, to cyber physical systems. Successful applications appear regularly. Another reason is that the POPLmark Challenge promoted the use of proof assistants in the programming language community. Consequently there is considerable interest in tightly integrating proof assistants with programming languages Brady (2011), Casinghino et al. (2014), Strub et al. (2012), Chlipala (2013). Already there are too many significant applications to cite even all the major successes. One of the most cited and most impressive examples is the creation of the CompCert C compiler back end Leroy (2006) with respect to a formal definition of C. There are many more very impressive examples. We note our recent work in synthesizing and deploying correct by construction distributed protocols, including Multi-Paxos Schiper et al. (2014).

Proof assistants are also used to formalize important mathematical results and for solving challenging open problems from mathematics. Coq was used to formalize major results such as the *Four Color Theorem* Gonthier (2008) and the *Odd Order Theorem* Gonthier et al. (2013). This task involved proving numerous important lemmas. Proof assistants are shaping and formalizing homotopy type theory as explored in *Homotopy Type Theory* Univalent Foundations Program (2013). The Minlog proof assistant is also used in important mathematical examples Berger and Schwichtenberg (1994). Reading these completely formal definitions, theorem, and proofs is a lesson in the value of formal precision and unsurpassed clarity. There are also impressive uses of proof assistants in education. The *Software Foundations* work is one of the best known education efforts Pierce et al. (2013), and there are other examples Chlipala (2013).

The Future of Proof Assistants The future of proof assistants is hard to predict in detail, but in broad outline, it now appears that reality will exceed the dreams of those who started the process – going back at least to the Boyer and Moore system and their book *Computational Logic* Boyer and Moore (1979) 36 years ago. Another early book on the subject was *Implementing Mathematics with Nuprl* Constable et al. (1986), and now we can imagine titles such as *Implementing Compilers with Coq*, *Implementing Real Analysis with HOL* (that is essentially Harrison's book Harrison (1998)), *Implementing Distributed Protocols with Nuprl*, *Implementing Constructive Euclidean Geometry with Nuprl*, and so forth Beeson (2012). It is conceivable that at some point, the whole of the book *First-Order Logic* will be automated by a proof assistant with respect to constructive evidence semantics Underwood (1993), Underwood (1994).[2]

Such books would represent the impact of this proof technology on the world as reflected in university curricula. Readers might think that the *major impact* will be *deployed software systems that they can trust*. On the other hand, the deepest impact is likely to be on how we know and learn. For scientists, *what it means to know will be redefined* by how computers mediate the digital information that we help them create. It is because of this possibility that we must balance complete formalism with clear intuitive foundations lest formalisms kill the intuitive meanings so essential to human understanding. Smullyan's important works are a guide for how we can preserve intuition and elegance when we formalize results.

[2] In the extreme, some articles claim that eventually "intelligent" machines will use these tools to create mathematics that is beyond the abilities of unaided humans Aron (2015).

2.2 Constructive Type Theory Background

2.2.1 *Implemented Type Theories*

Four of the proof assistants mentioned above, Agda Bove et al. (2009), Coq Bertot and Castéran (2004), Nuprl Allen et al. (2006), and MetaPRL Hickey et al. (2003) implement versions of constructive type theories and generalize work of logicians on intuitionistic type theory Martin-Löf (1982), Martin-Löf (1998) and constructive set theory Aczel (1982), Friedman (1977). From the beginning Nuprl implemented Constructive Type Theory (CTT) Constable et al. (1986), a significant enrichment of Martin-Löf's *extensional type theory* from 1982 Martin-Löf (1982), call it ML82. CTT provides a precise foundational semantics Allen (1987) that justified the addition of several new types to constructive type theory, including recursive types, quotient types, refinement types, and *partial types* (in which elements can diverge, as in programming language types).

One of the most powerful type constructors added to Nuprl after 2000 is the *intersection type* and the intersection of a family of types $\bigcap_{i:A} B(i)$. These are the elements that belong to all of the types $B(i)$. This constructor becomes a *uniform universal quantifier* in logic and is the key to defining a powerful new concept of *uniform validity* used to give a new *completeness theorem* for first-order logic Constable and Bickford (2014). Applications show this to be one of the most useful types for specification. This type is used to formalize the Smullyan result on Boolean evaluation. The other new types such as *dependent intersection* Kopylov (2003), *objects* Kopylov (2004), and *union* are also interesting, but we do not have occasion to expand on them in this article.

The *partial types* Constable and Smith (1993) allow Nuprl to write *partial recursive functions* over any data type, thus making it a universal programming language, unlike ML82, Agda and Coq which only support total types and total functions on them. CTT begins with total types T and defines the corresponding partial types $\overline{T}$ as including possibly diverging terms with the property that if they converge, they belong to the type T. Every partial type is inhabited by any term that is known to diverge, such as $fix(\lambda(x.x))$. We give this term the short name $\perp$. We can take fix to be the **Y** combinator,

$$\mathbf{Y} = \lambda f(\lambda x(f(xx)))(\lambda x(f(xx))).$$

It has the typing rule needed here, and we use it later in the article as well.

Like the **Y** combinator, the fix operator is used to define recursive functions. For example the recursive multiplication function

$$mult(x, y) = if\ y = 0\ then\ 0\ else\ mult(x, y - 1) + x$$

is written as

$$fix(\lambda(mult.\lambda(x, y.if\ y = 0\ then\ 0\ else\ mult(x, y-1) + x))).$$

To evaluate the above term, we substitute the entire fix expression for $mult$ in the definition. This is a very clear definition of recursion in the style of the **Y** combinator.

The simplest form of the typing rule we require later is this. If f has type $\overline{T} \to \overline{T}$, then $fix(f)$ has type $\overline{T}$. A common example is a functional F of type

$$(\overline{T} \to \overline{T}) \to (\overline{T} \to \overline{T}).$$

The least fixed point of this functional is a partial function in the type $(\overline{T} \to \overline{T})$. It is given by the fix operator. In constructive type theory the functionals and functions are effectively computable.

2.3 Validating Hoare Logic using Partial Types

We start this section by looking briefly at how partial types can naturally express many basic ideas of computability theory Constable and Smith (1993). For instance, the halting problem can be stated in a particularly clear way and it's unsolvability given an elegant short proof, the kind Smullyan produced in so many of his writings. The type $\mathbb{B}$ is the type of Booleans.

Unsolvability results using partial types Here is a statement of the unsolvability of the halting problem in terms of partial types and a proof of the result.

Given any non-empty type T, there is no function $h : \overline{T} \to \mathbb{B}$ such that $\forall x : \overline{T}.(h(x) = tt \text{ iff } x \downarrow)$, e.g. no internal computable h solves the halting problem on $\overline{T}$.

Consider the special case of $T = N$ for concreteness; the proof works verbatim for any non-empty type T. Note that in constructive type theory, all functions are effectively computable. So when we say that no halting detector exists, we are saying that no internally computable function can solve the halting problem. Since the Nuprl type theory includes all partial recursive functions and is thus universal, the result we prove is a faithful version of the halting problem as it is widely known.

Theorem $\neg\exists h : \overline{\mathbb{N}} \to \mathbb{N}.\forall x : \overline{\mathbb{N}}.(h(x) = 0 \text{ iff } x \downarrow)$

Proof Suppose such an $h : \overline{\mathbb{N}} \to \mathbb{N}$ could be constructed in this theory, then consider the function $\lambda(x.zero(h(x); \bot; 0)$, where the $zero$ function tests its first argument for having the value zero and returns $\bot$ in that case and 0 otherwise. Its type is $\overline{\mathbb{N}} \to \overline{\mathbb{N}}$. So $fix(\lambda x.zero(h(x); \bot; 0))$ has type $\overline{\mathbb{N}}$. Let d be this value in $\overline{\mathbb{N}}$.

So $fix(f) = d \in \overline{\mathbb{N}}$ by the fixed point typing rule. Now consider the value of $h(d)$. The value belongs to $\mathbb{N}$, so we have two cases, $(h(d) = 0) \lor (h(d) \neq 0)$. We show that neither case is possible, hence we have no such h.

1. If $h(d) = 0$ then $zero(h(d); \bot; 0) = \bot$, contrary to the defining requirement on h that when $h(d) = 0$, the function should return 0.

2. If $h(d) \neq 0$, then $zero(h(d); \bot; 0) = 0$, by the definition of h, the value should be $\bot$, a diverging element. □

2.3.1 Hoare Logic for Programs

Hoare's logic of *partial correctness* is based on a relation between assertions and commands written as follows: $\{A\}C\{B\}$ with the meaning that if assertion A is true of state s before command C is executed on s, and if C terminates in state s', then assertion B is true in state s'. Using partial correctness arguments, it is easy to explain some of the simple rules for proving properties of programs. For example, Hoare's famous *while rule* explains the important notion of a *loop invariant*. Here is the simple rule for assignment statements (commands to update the value of a variable). The *assignment command* $X := exp$ executes as follows. First the expression exp is evaluated in the current *state* of the program. The state is simply a mapping from the program variables, say $X, Y, Z, X_1, Y_1, Z_1, \ldots$ to values. Next, the state is modified by assigning the value of exp just computed to the variable X. Suppose the values are natural numbers, then a typical assignment command could be $X := 2 \times (X + 1)$ which adds one to the value of X, then doubles it and assigns that new value to X.

To prove something about the result, say that if before executing this command the value of X is non-negative, then after this operation X is a positive even number, we only need to know that the value of X when exp is evaluated is non-negative. Thus the following Hoare style assertion about the command is valid where $EvenPositive(y)$ is the assertion that y is an even positive number.

$$\{EvenPositive(2 \times (X + 1))\}\ X := 2 \times (X + 1)\ \{EvenPositive(X)\}.$$

A rule proposed for functions in Hoare logic was this. If $\{A(x)\}P\{B(y)\}$, and the function f on input x executes the program P and returns its value in variable y, then we can conclude $\forall x.(A(x) \Rightarrow B(f(x)/y))$.

If P is a command that never terminates, say because it is executing an unbound loop such as $while\ true\ do\ skip\ od$, then we know $\{True\}P\{False\}$. So from the proposed function rule we would know that $\forall x.(True \Rightarrow False(f(n)/x))$, leading directly to a contradiction.

In constructive type theory, the notion of a partial type $\overline{T}$ is defined as the type of all expressions which have type T if they terminate. So given a recursive function defined as $f(x) = if x = 0\, then\, 0\, else\, f(x + 1)$, we know that $f(1)$ will have the type $\overline{\mathbb{N}}$.

If we state the Hoare rule for *Loop* in the form $\{True\}Loop\{\overline{False}\}$, then the above argument does not lead to a contradiction because $\{\overline{False}\}$ does not assert $False$. All we know is that there is an expression which does not terminate that would be evidence for $False$ if it did terminate. In this form the idea seems trivial, but in the context of a program that includes assertions and in which the logical rules about assertions demonstrate effectively how to build evidence for the assertions as the program executes, the use of partial types for assertions conveys very precisely the computational meaning of the hybrid object called an *asserted program*, that is an imperative program annotated with assertions.

2.4 Formalizing a Theorem of Smullyan from *First-Order Logic*

In this section we look at a specific theorem from *First-Order Logic* that has an elegant implicitly constructive proof. In Chapter I Sect. 2.2 on Boolean Valuations, Smullyan shows how to extend any interpretation v_0 for *all* of the propositional variables to *at most one Boolean valuation* of the set of all formulas of propositional logic. He denotes this set of all formulas by E; we will use F for Formulas. He does not prove this by giving a recursive definition of an evaluation function with input e_0 for the values of the variables of a specific formula f being evaluated, which is the way it would be done in a typical computer science course on logic. Instead he gives a constructive proof that there is a unique Boolean valuation for the entire set (or type) F. From this constructive existence proof, *we can extract the recursive evaluation function* for formulas f of the type F. That is, we can build the normal recursive evaluation function for formulas from Smullyan's existence proof for of a valuation for the entire type F.

A *Boolean valuation* (page 10) is a function v from the type of all formulas of propositional logic, F into the Booleans such that four conditions are met, B1 to B4. B1 says that $\neg X$ receives the opposite value of that assigned to X. B2 defines the truth value for &, B3 for $\vee$, and B4 for implication, $\Rightarrow$.

Examining this small example illustrates an important idea in the approach to program specification using type theory. The natural approach to defining the evaluation function *eval* for Boolean formulas, say formula f in the type of all Boolean formulas, F, is to supply two arguments to *eval*, first the formula f and then an assignment of values to the variables occurring in f, say $eval(f, v)$ where function v maps the propositional variables of f, say $var(f)$ into Booleans, i.e. $v : var(f) \to \mathbb{B}$. So the type of the evaluation function is:

$$eval : (f : F \to v : (var(f) \to \mathbb{B}) \to \mathbb{B}.$$

The type expression for *eval* is a *dependent function type* because the type definition *depends* on the first input to *eval*, namely the formula f being evaluated. That formula is used to pick out the collection of variables the actually occur in f. This type could simply be a list of variables, say the type $VarList$. This typing matches the truth table approach to evaluating a boolean function. Given the formula f, we list all the variables of f as columns names for the truth table. Each row of the table gives the value of the formula for a particular assignment of truth values to the propositional variables.

Until very recently, no programming language would allow the type definition given above for the evaluation function because *dependent function types* were not allowed. These types are common in the type theories of the proof assistants mentioned above. We give the exact Nuprl types below. Smullyan deals with this issue by using the type of all possible truth assignments, v_0, to the unbounded list of all legal propositional variables. This is not a notion that can be efficiently implemented, and he does not attempt that. Instead he gives an "infinitary" account of evaluation in the precise mathematics, and then suggests how this account relates to a practical method of evaluation given in parentheses. His evaluation result has this form:

$$\forall v_0 : (Var \rightarrow \mathbb{B}).\forall f : F.\exists v : BooleanValuation(f, v).$$

He then proves that this Boolean valuation is unique. We want to show how to obtain a computable evaluation procedure from his proof of the existence and uniqueness of a $BooleanValuation$.

The methodology of extracting an algorithm from a constructive proof or from one that can easily be made constructive is supported by proof assistants for constructive mathematics mentioned above, Agda, Coq, Nuprl, and MetaPRL. They all implement the idea that from constructive existence proofs, one can build the object claimed to exist. Smullyan's result is more general than the standard recursive *evaluation function* whose first input is a formula f and whose other input is a valuation only for the variables of f. This standard approach can be handled using the dependent function type as we just mentioned above. Smullyan could have used this approach in his informal set theoretic meta language, but he choose to present the approach where one starts with an assignment to all propositional variables. This approach happens to be more challenging to formalize in type theory.

Our first efforts to formalize this theorem used a concept called a "very dependent function type" to handle the construction implicit in Smullyan's account, one that does not proceed by looking only at the variables in the formula being assigned a value. We eventually found that a very elegant account could be given using the *intersection type*. The type used by Nuprl is related to earlier work in logic Pottinger (1980), Coppo and Dezani-Ciancaglini (1978). This idea is presented in detail below in the subsection on constructing valuations.

Here is what we said in the first unpublished technical report that we wrote on this topic with the informative title *Expressing and Implementing the Computational Content Implicit in Smullyan's Account of Boolean Valuations* Allen et al. (2004):

"In Smullyan's classic book, *First-Order Logic*, the notion of a Boolean valuation is central in motivating his analytical tableau proof system. Smullyan shows that these valuations are unique if they exist, and then he sketches an existence proof. In addition he suggests a possible computational procedure for finding a Boolean valuation, but it is not related to to the existence proof. A computer scientist would like to see the obvious explicit recursive algorithm for evaluating propositional formulas and a demonstration that the algorithm has the properties of a Boolean valuation. Ideally, the algorithm would be derived from the existence proof. It turns out to be unexpectedly difficult to find a natural existence proof from which the algorithm can be extracted, and moreover, the implicit computational content of Smullyan's argument is not found where one might expect it. We show that using the notion of a very dependent function type, it is possible to specify the Boolean valuation and prove its existence constructively so that the natural recursive algorithm is extracted and is known to have the mathematically required properties by virtue of its construction. We illustrate all of these points using the Nuprl proof development system."

The new approach summarized here from the article on polymorphic logic Bickford and Constable (2012) does not use the very dependent function type. Instead we use the intersection type.

2.4.1 *Uniform Universal Quantification*

The standard universal quantifier $\forall x : T.\ P(x)$ is defined to be the *dependent function type*, $x : T \rightarrow P(x)$. A witness $f \in \forall x : T.\ P(x)$ is a function $f \in x : T \rightarrow P(x)$ that maps any $x \in T$ to evidence for $P(x)$. There are cases where a single p is a uniform witness for $P(x)$ for any $x \in T$. In this case, p is a member of the *intersection type*, $\bigcap_{x:T} P(x)$. Such a p is not a function with input $x \in T$, instead it is a witness for $P(x)$, *polymorphic* or *uniform* over all $x \in T$.

The polymorphic universal quantifier $\forall[x : T].\ P(x)$ is defined to be $\bigcap_{x:T} P(x)$. The brackets around the bound variable indicate that the witness does not "use" the parameter x. Classically, $\forall x : T.\ P(x)$ and $\forall[x : T].\ P(x)$ have the same meaning, but constructively they do not. Evidence p for the proposition with the polymorphic quantifier is likely to be more efficient since it does not need to be given an input $x \in T$. $\mathbb{P}$ is the type of propositions of type theory. A "false" proposition is an empty type, so it is extensionally equal to $Void$. A "true" proposition is one for which there is evidence in the type that captures its meaning, so that its meaning is a non-empty type. The members of the type are the reasons for knowing the proposition.

2.4.2 *Rules for* $\forall[x : T].\ P(x)$

The rules for proving $\forall[x : T].\ P(x)$ are the rules for inhabiting $\bigcap_{x:T} P(x)$. These use contexts with *hidden declarations*. To prove $\Gamma \vdash \bigcap_{x:T} P(x)$ we prove $\Gamma, [x : T] \vdash P(x)$. The brackets on the declaration $[x : T]$ added to the context Γ indicate that the evidence for T is hidden.

To prove this sequent, we use rules for proving $P(x)$, and no rules use hidden declarations. The hidden declarations are automatically unhidden once the sequent is refined to a conclusion of the form $t_1 = t_2 \in T$, since those rules all extract a fixed witness term Ax (because we consider equality propositions to have no constructive content) the extract of any proof of $\Gamma, [x : T] \vdash P(x)$ will not include the hidden parameter x. In particular, the proposition $t \in T$ is simply an abbreviation for $t = t \in T$, so when proving a typing judgement, the hidden declarations are unhidden.

2.4.3 A Uniform Complete Induction Principle

The principle of *complete induction* over the natural numbers, $\mathbb{N}$, can be written in type theory as

$$\forall P : \mathbb{N} \to \mathbb{P}. \, (\forall n : \mathbb{N}. \, (\forall m : \mathbb{N}_n. \, P(m)) \Rightarrow P(n)) \Rightarrow (\forall n : \mathbb{N}. \, P(n)).$$

The type $\mathbb{N}_n$ is the set type $\{m : \mathbb{N} \mid m < n\}$ whose members are the natural numbers less than n. A witness for the induction principle is a member *Ind* of the corresponding dependent function type

$$P : (\mathbb{N} \to \mathbb{P}) \to (n : \mathbb{N} \to (m : \mathbb{N}_n \to P(m)) \to P(n)) \to (n : \mathbb{N} \to P(n)).$$

The witness *Ind* has the form $\lambda P. \, \lambda G. \, \lambda n. \, \ldots$. It takes inputs P, G, and n, where G has type $(n : \mathbb{N} \to (m : \mathbb{N}_n \to P(m)) \to P(n))$, and produces a witness, $Ind(P, G, n)$,for $P(n)$.

If we restate the induction principle using the polymorphic universal quantifier, we get

$$\forall[P : \mathbb{N} \to \mathbb{P}]. \, (\forall[n : \mathbb{N}]. \, (\forall[m : \mathbb{N}_n]. \, P(m)) \Rightarrow P(n)) \Rightarrow (\forall[n : \mathbb{N}]. \, P(n)).$$

Proving this is equivalent to the construction of a witness W of type

$$\bigcap_{P : (\mathbb{N} \to \mathbb{P})} (\bigcap_{n : \mathbb{N}} (\bigcap_{m : \mathbb{N}_n} P(m)) \to P(n)) \to (\bigcap_{n : \mathbb{N}} P(n)).$$

W will have the form $\lambda F. \ldots$ and take an $F \in (\bigcap_{n : \mathbb{N}} (\bigcap_{m : \mathbb{N}_n} P(m)) \to P(n))$ and produce a member, $W(F)$, of $(\bigcap_{n : \mathbb{N}} P(n))$. The input F is a function that takes an $x \in (\bigcap_{m : \mathbb{N}_n} P(m))$ and produces a witness, $F(x)$ for $P(n)$. The result $W(F)$ is a uniform witness for all the $P(n)$, $n \in \mathbb{N}$. Such a W appears to be a fixed point operator, and we can, in fact, prove the polymorphic induction principle using any fixed point combinator **fix** that satisfies

$$\mathbf{fix}(F) \sim F(\mathbf{fix}(F)).$$

A technical point is that the relation $\sim$ is the symmetric-transitive closure of $\mapsto$, where $t_1 \mapsto t_2$ if a single primitive computation step such as β-reduction, expanding definitions (δ-reduction), or reducing another primitive ($+$, $*$, $\ldots$, on numbers, projections on pairs, etc.) transforms t_1 into t_2. In computational type theory all types are closed under $\sim$, so we have *subject reduction* :

$$x \in T, x \sim y \vdash y \in T.$$

Lemma 1

$$\forall[P\colon \mathbb{N} \to \mathbb{P}].\ (\forall[n\colon \mathbb{N}].\ (\forall[m\colon \mathbb{N}_n].\ P(m)) \Rightarrow P(n)) \Rightarrow (\forall[n\colon \mathbb{N}].\ P(n)).$$

Proof Given $[P \in \mathbb{N} \to \mathbb{P}]$ and $f : \forall[n\colon \mathbb{N}].\ (\forall[m\colon \mathbb{N}_n].\ P(m)) \Rightarrow P(n)$ we must construct a member of $(\forall[n\colon \mathbb{N}].\ P(n))$ (without using P). We show that $\mathbf{fix}(f)$ (which is independent of P) is in $(\forall[n\colon \mathbb{N}].\ P(n))$. Since this is a proof of a typing judgement, we may now use the declarations that were formerly hidden. This is a subtle point about the proof rules for set types. In this case it is clear from the intuitions behind the proof.

Let Γ be the context $P : \mathbb{N} \to \mathbb{P},\ f : \bigcap_{n\colon \mathbb{N}}(\bigcap_{m\colon \mathbb{N}_n} P(m)) \Rightarrow P(n)$. We must show $\Gamma,\ n : \mathbb{N} \vdash \mathbf{fix}(f) \in P(n)$ and we use the complete induction principle on n. Thus, we show that $\mathbf{fix}(f) \in P(n)$ follows from the assumptions

$$\Gamma, n : \mathbb{N}, \forall m\colon \mathbb{N}_n.\ \mathbf{fix}(f) \in P(m).$$

But this implies $\mathbf{fix}(f) \in (\bigcap_{m\colon \mathbb{N}_n} P(m))$, and therefore, using the polymorphic type of f, $f(\mathbf{fix}(f)) \in P(n)$. Since $f(\mathbf{fix}(f)) \sim \mathbf{fix}(f)$, we have $\mathbf{fix}(f) \in P(n)$. □

We carried out this proof in Nuprl using for the fixed point combinator the Y-combinator,

$$Y = \lambda f(\lambda x(f(xx)))(\lambda x(f(xx))).$$

The extract of the proof, computed by the system, is simply the term Y.

2.4.4 Constructing Smullyan's Valuations

Smullyan's *First-Order Logic* Smullyan (1968) starts with a definition of propositional logic. A key concept is the notion of a *valuation* of a propositional formula given an assignment of truth values to its propositional variables. Smullyan constructively proves the existence and uniqueness of valuations. We create valuations by extracting the computational content of a proof as we illustrate below. The existence of Boolean valuations is expressed in the normal way for $\mathbb{B}$ the Boolean values.

$$\forall x\colon \mathit{Form}.\ \forall v_0\colon \mathit{Var}(x) \to \mathbb{B}.\ \exists f\colon \mathit{Sub}(x) \to \mathbb{B}.\ \mathit{valuation}(x, v_0, f)$$

The formal definitions of formulas, $Form$ and sub-formulas, Sub, are straightforward. We define a standard type for the formulas of propositional logic with the following recursive type:

$$Form := var(Atom) \mid \textbf{not}\ (Form) \mid Form\ \textbf{and}\ Form \mid Form\ \textbf{or}\ Form \mid Form\ \textbf{implies}\ Form.$$

From this formal definition, the proof assistant automatically creates the recursive type *Form* together with constructors, destructors, and recognizers for each case, as well as an induction principle and an induction operator witnessing the induction principle. This technology is common in all the proof assistants for type theory such as Agda, Coq, Nuprl, and MetaPRL. In particular, we have a predicate $var?(x)$ that recognizes variables.

Using the induction operator we define the sub-formula relation $q \subseteq p$ on formulas, and show that it is reflexive and transitive. The types $Var(x)$ and $Sub(x)$ are then defined using the CTT set type:

$$Sub(x) = \{v : Form \mid v \subseteq x\}$$
$$Var(x) = \{v : Form \mid v \subseteq x \ \wedge \ var?(v)\}$$

The induction operator for the type can also be treated non-inductively as a simple case operator. We use this to define the value of a formula p given an assignment v_0 and a function g defined on the proper sub-formulas of p. This is the key definition corresponding to Smullyan's definition at the top of page 11 in the Dover edition of his book.

$$\begin{aligned} extend(v_0, g, p) = \ & case(p) \\ & var(v) \Rightarrow v_0(v) \\ & \textbf{not}\ q \Rightarrow bnot(g(q)) \\ & q_1\ \textbf{and}\ q_2 \Rightarrow band(g(q_1), g(q_2)) \\ & q_1\ \textbf{or}\ q_2 \Rightarrow bor(g(q_1), g(q_2)) \\ & q_1\ \textbf{implies}\ q_2 \Rightarrow bimp(g(q_1), g(q_2)) \end{aligned}$$

Here, $bnot$, $band$, bor, and $bimp$ are the obvious Boolean functions defined on $\mathbb{B}$, the Boolean values.

For a function f of type $Sub(x) \to \mathbb{B}$ to be a valuation of x given the assignment v_0 of type $Var(x) \to \mathbb{B}$ it must satisfy the constraint $valuation(x, v_0, f)$ defined by

$$valuation(x, v_0, f) \Leftrightarrow \forall p : Sub(x).\ f(p) = extend(v_0, f, p).$$

This defines a valuation as a function that *correctly extends itself.* This captures the elegant idea in Smullyan's proof at the top of page 11 of his *First-Order Logic book.* To quote him exactly: "Thus we have now defined what it means for a formula X to be true under an *interpretation.*"

Lemma 2

$$\forall x: Form. \forall v_0: Var(x) \to \mathbb{B}. \exists f: Sub(x) \to \mathbb{B}. valuation(x, v_0, f)$$

Proof The induction operator on formulas provides a rank function $|x|$ with range $\mathbb{N}$. It decreases on proper sub-formulas and assigns variables rank 0. We define next a bounded valuation by

$$bddval(n, x, v_0, f) \Leftrightarrow \forall p: Sub(x). |x| < n \Rightarrow f(p) = extend(v_0, f, p).$$

Given the context $\Gamma = x : form, v_0 : Var(x) \to \mathbb{B}$ it must be shown that $\Gamma \vdash \exists f : . Sub(x) \to \mathbb{B} valuation(x, v_0, f)$. The "cut" rule asserts the (polymorphically quantified)

$$\forall [n: \mathbb{N}]. \exists f: Sub(x) \to \mathbb{B}. bddval(n, x, v_0, f).$$

From this assertion we easily complete the proof by choosing n to be $|x| + 1$. To prove the assertion we invoke the induction principle in Lemma 1.

Lastly we must prove that from $n : \mathbb{N}$, $\bigcap_{m: \mathbb{N}_n} \exists f : Sub(x) \to \mathbb{B}. bddval(m, x, v_0, f)$ it follows that $\exists f: Sub(x) \to \mathbb{B}. bddval(n, x, v_0, f)$. To accomplish this we let f be a member of the type in the induction hypothesis, and then use $\lambda p.extend(v_0, f, p)$. □

This existence proof is essentially the one given by Smullyan Smullyan (1968) pages 10 and 11. It was carried out by Mark Bickford using Nuprl, and the extract of the lemma, constructed by Nuprl, is the term

$$\lambda v_0.(Y(\lambda f, p.extend(v_0, f, p))).$$

These results go beyond the unpublished work Allen et al. (2004) that used very dependent functions to find the computational content in Smullyan's account of Boolean evaluation. That article *Expressing and Implementing the Computational Content Implicit in Smullyan's Account of Boolean Valuations*, quoted earlier is available at Publications on the PRL project page, www.nuprl.org.

References

Aczel, P. (1982). The type theoretic interpretation of constructive set theory: Choice principles. In S.S. Troelstra & D. van Dalen, (Eds.), *The L.E.J. Brouwer Centenary Symposium*. North Holland.

Allen, S. F. (June 1987). A Non-type-theoretic Definition of Martin-Löf's Types. In D. Gries (Ed.) *Proceedings of the 2nd IEEE Symposium on Logic in Computer Science* (pp. 215–224). IEEE Computer Society Press.

Allen, S. F., Constable, R. L. & Fluet, M. (2004). Expressing and implementing the computational content implicit in Smullyan's account of Boolean valuations. Technical Report TR2004-1933, Department Computer Science, Cornell University

Allen, S., Bickford, M., Constable, R., Eaton, R., Kreitz, C., Lorigo, L., et al. (2006). Innovations in computational type theory using Nuprl. *Journal of Applied Logic*, *4*(4), 428–469.
Aron J. (2015). Beyond knowledge. *New Scientist*, pp. 28–31.
Basin, D. & Wolff, B. (Eds.) (2003). *Proceedings of the* 16th *International Conference on Theorem Proving in Higher Order Logics (TPHOLs 2003)*, volume 2758 of *Lecture Notes in Computer Science*. Springer-Verlag.
Beeson, M. J. (2012). Logic of ruler and compass constructions. In S. B. Cooper, A. Dawar, & B. Loewe, (Eds.), *Computability in Europe 2012*, Lecture Notes in Computer Science (pp. 46–55). Springer, 2012.
Benl, H., Berger, U., Schwichtenberg, H., et al. (1998). Proof theory at work: Program development in the Minlog system. In W. Bibel & P. G. Schmitt (Eds.), *Automated Deduction* vol. II. Kluwer.
Berger, U., & Schwichtenberg, H. (1994). Program extraction from classical proofs. In D. Leivant (Ed.), *Logic and Computational Complexity* (pp. 77–97). Berlin: Springer.
Bertot, Y., & Castéran, P. (2004). *Interactive Theorem Proving and Program Development; Coq'Art: The Calculus of Inductive Constructions*. Texts in Theoretical Computer Science: Springer-Verlag.
Bickford, M., & Constable, R. L. (2012). *Polymorphic logic*. In Logic, Construction, Computation: Ontos Verlag, Festschrift for Helmut Schwichtenberg.
Bove, A., Dybjer, P., Norell, U. (2009). A brief overview of Agda – a functional language with dependent types. In Stefan Berghofer, Tobias Nipkow, Christian Urban, and Makarius Wenzel, editors, *LNCS 5674, Theorem Proving in Higher Order Logics* (pp. 73–78). Springer.
Boyer, R. S., & Moore, J. S. (1979). *A Computational Logic*. New York: Academic Press.
Brady, E. (2011). Idris:systems programming meets full dependent types. In *Programming Languages meets Program Verification, PLPV 2011* (pp. 43–54). ACM.
Casinghino, C., Sjberg, V., & Weirich, S. (2014) Combining proofs and programs in a dependently typed language. In *41st ACM SIGPLAN-SIGACT Symposium on Principles of Programming Languages* (pp. 33 – 55). San Diego, CA.
Chlipala, A. (2013). *Certified Programming with Dependent Types*. Cambridge, MA: MIT Press.
Constable, R. L., Allen, S. F., Bromley, H. M., Cleaveland, W. R., Cremer, J. F., Harper, R. W., et al. (1986). *Implementing Mathematics with the* `Nuprl` *Proof Development System*. NJ: Prentice-Hall.
Constable, R., & Bickford, M. (2014). Intuitionistic Completeness of First-Order Logic. *Annals of Pure and Applied Logic*, *165*(1), 164–198.
Constable, R. L., & Smith, S. F. (1993). Computational foundations of basic recursive function theory. *Journal of Theoretical Computer Science*, *121*(1&2), 89–112.
Coppo, M., & Dezani-Ciancaglini, M. (1978). A new type assignment for λ-terms. *Archive for Mathematical Logic*, *19*(1), 139–156.
Crary, K. (August 1998). *Type–Theoretic Methodology for Practical Programming Languages*. Ph.D. thesis, Cornell University, Ithaca, NY.
Fitting, M. (1969). *Intuitionistic model theory and forcing*. Amsterdam: North-Holland.
Friedman, H. (1977). Set theoretic foundations for constructive analysis. *Annals of Math*, *105*, 1–28.
Gonthier, G. (2008). Formal proof - the Four Color Theorem. *Notices of the American Math Society*, *55*, 1382–1392.
Gonthier, G., Asperti, A., Avigad, J., Bertot, Y., Cohen, C., Garillot, F., et al. (2013). A Machine-checked Proof of the Odd Order Theorem. In S. Blazy, C. Paulin-Mohring, & D. Pichardie (Eds.), *ITP* (Vol. 7998)., Lecture Notes in Computer Science, (pp. 163–179) Springer.
Gordon, M., & Melham, T. (1993). *Introduction to HOL: A theorem proving environment for higher-order logic*. Cambridge: Cambridge University Press.
Harrison, J. (1998) *Theorem proving with the real numbers*. Springer-Verlag.
Hickey, J., Nogin, A., Kopylov, A., et al. (2003) `MetaPRL` — A Modular Logical Environment. In Basin and Wolff [6], pp. 287–303.
Hoare, C. A. R. (1971). Procedures and parameters: An axiomatic approach. *Semantics Algorithm. Lang.*, *188*, 102–116.

Kopylov, A. (2003) Dependent intersection: A new way of defining records in type theory. In *Proceedings of* 18th *IEEE Symposium on Logic in Computer Science*, pp. 86–95.
Kopylov, A. *Type theoretical foundations for data structures, classes, and objects*. Ph.D. thesis, Cornell University, Ithaca, NY, 2004.
Leroy, X. (2006) Formal certification of a compiler back-end or: programming a compiler with a proof assistant. In *Proceedings of the 33d ACM SIGPLAN-SIGACT symposium on Principles of programming languages* (pp. 42–54). ACM Press.
Martin-Löf, P. (1982). Constructive mathematics and computer programming. In *Proceedings of the Sixth International Congress for Logic, Methodology, and Philosophy of Science* (pp. 153–175). Amsterdam: North Holland.
Martin-Löf, P. (1998) An intuitionistic theory of types. In G. Sambin & Smith, J.M. (Eds.) *Twenty-Five Years of Constructive Type Theory*, volume 36 of *Oxford Logic Guides*, (pp. 127–172). Oxford: Clarendon Press.
Nerode, A., & Shore, R. (1994). *Logic for Applications*. New York: Springer-Verlag.
O'Donnell, M. J. (1982). A critique of the foundations of Hoare style programming logics. *Commications of the Association of Computing Machinery*, *25*(12), 927–935.
Pierce, B. C., Casinghino, C., Greenberg, M., Sjberg, V., & Yorgey B (2013). Software Foundations. Electronic.
Pottinger, G. (1980) A type assignment for the strongly normalizable λ-terms. In J. P. Seldin & J. R. Hindley, (Eds.), *To H. B. Curry: Essays in Combinatory Logic, Lambda Calculus and Formalism* (pp. 561–577). London: Academic Press.
Raymond, M. (1961). *Smullyan*. Theory of Formal Systems: Princeton University Press.
Raymond, M. (1987). Smullyan. Quotation and self-reference. In S. J. Bartlett & P. Suber (Eds.), *Self-Reference: Reflections on Reflexivity*. Dordrecht, Holland: Martinus Nijhoff.
Raymond, M. (1992). *Smullyan*. New York: Gödel's Incompleteness Theorems. Oxford University Press.
Scott, D. (1976). Data types as lattices. *SIAM J. Comput.*, *5*, 522–587.
Schiper, N., Rahli, V., Van Renesse, R., Bickford, M., & Constable, R. L. (2014). Developing correctly replicated databases using formal tools. In *DSN 2014: The 44th Annual IEEE/IFIP International Conference on Dependable Systems and Networks*.
Smullyan, R. M. (1968). *First-order logic*. New York: Springer-Verlag.
Smullyan, R. M. (1983). *5000 B.C. and Other Philosophical Fantasies*. NY: St. Martin's Press.
Smullyan, R. M. (1984). *What is the name of this book?* Pelican.
Smullyan, R. M. (1994). *Diagonalization and self-reference*. Number 27 in Oxford Logic Guides. Oxford: Clarendon Press.
Strub, P. -Y., Swamy, N., Fournet, C. & Chen, J. (2012). Self-certification: Bootstrapping certified typecheckers in F star with Coq. In *Proceedings of the ACM Symposium on Principles on Programming Languages (POPL'12)* (pp. 571– 583). ACM.
Underwood, J. L. (1993) The tableau algorithm for intuitionistic propositional calculus as a constructive completeness proof. In *Proceedings of the Workshop on Theorem Proving with Analytic Tableaux, Marseille, France*, (pp. 245–248). Available as Technical Report MPI-I-93-213 Max-Planck-Institut für Informatik, Saarbrücken, Germany.
Underwood, J. L. (1994) Aspects of the computational content of proofs. Department of Computer Science TR94-1460, Cornell University, Ithaca, NY, October
Univalent Foundations Program. *Homotopy Type Theory*. Univalent Foundations Program, 2013.

Chapter 3
Adaptive Fault Diagnosis using Self-Referential Reasoning

Robert Cowen

Abstract The problem is to determine which processors are reliable in a remote location by asking"Yes or No" questions. The processors are of three types: those that always tell the truth, those that always lie, and those the sometimes tell the truth and sometimes lie. Using self-referential reasoning, along with earlier techniques, we can regard both the truth-tellers and liars as reliable and thus the tackle situations when fewer than half the processors are truth-tellers.

3.1 Introduction

I first encountered logical puzzles about "Knights," who always tell the truth, "Knaves," who always lie, and "Normals," who sometimes tell the truth and sometimes lie, in Raymond Smullyan's wonderful book Smullyan (1978), "What is the Name of this Book?" In the book, Smullyan not only entertains, but takes the reader on a journey, visiting many bizarre lands and finally ending in the Gödelian Islands giving a lucid explanation of Gödel's famous Incompleteness Theorem. The reasoning employed is always enlightening as Smullyan reaches for more general principles; for example, on page 110, it is demonstrated that if a Knight or Knave proclaims, "If I am a Knight, then P," where P is any proposition, then the speaker must be a Knight and P must be true! This fact enables one to more easily solve several of the puzzles. Moreover, it is a simple example of "self-referential reasoning,"where the speaker makes a statement that references himself. In this paper, we shall apply this kind of reasoning to identify faulty computers.

We wish to determine which processors (components, chips, etc.) are reliable in a remote location using as few "Yes or No" questions as possible. It is assumed that the processors have the ability to test each other. The questions will be suggested by us, but asked by the processors, themselves, to each other. It may be that it is too dangerous for us to visit this remote location, in person, to test the processors; for example, we might need to test the reliability of a system in outer space or inside

R. Cowen
Queens College, CUNY, New York, USA
e-mail: robert.cowen@gmail.com

M. Fitting and B. Rayman (eds.), *Raymond Smullyan on Self Reference*,
Outstanding Contributions to Logic 14, https://doi.org/10.1007/978-3-319-68732-2_3

a nuclear reactor. It is further assumed that the majority of procesors are reliable. "Reliable" is often taken to mean "tells the truth;" however, in this paper, we shall take it to mean we are able to obtain correct information from it. We will explain this distinction below. This problem of determining which components are reliable by questions the components ask themselves about each other has a long history (see, for example, Alonso et al. 2004; Blecher 1983; Hakimi and Nakajima 1984; Hakimi and Shmeichel 1984; Wildon 2010).

In this paper, the processors are assumed to be of three types which we term, *à la* Smullyan (1978): Knights, Knaves and Normals. "Knights" always tell the truth, "Knaves" always lie, and "Normals" sometimes tell the truth and sometimes lie. Using self-referential reasoning, we shall show how to find all the reliable ones, by asking only "Yes or No" questions, assuming that the Knights and Knaves constitute a majority, that is, the Normals make up less than half of the total number of processors. In fact, by using appropriately constructed self-referential questions, *both* the Knights and the Knaves will give us the same valid answers, that is, both the Knights and Knaves can be considered to be "reliable." This contrasts with the usual approach, where the processors are asked directly, "Is that processor a Knight?" and it is assumed that the Knights constitute a majority of the processors (see Alonso et al. 2004; Blecher 1983; Wildon 2010).

Another way to look at the difference between our self-referential reasoning approach and the standard one is the following. In our approach, the interpretation of the word "reliable" is expanded. Instead of regarding just the Knights to be reliable and Knaves and Normals, unreliable, we consider *both* the Knights and Knaves to be reliable and only the Normals, to be unreliable. Thus, the sufficient condition, "the number of unreliables must be less than half the total number of processors," is more easily satisfied. For example, if we know there are 3 processors of each type, it is necessary to use self-referential questioning to find a reliable individual.

Our first algorithm finds a reliable individual, that is a Knight or a Knave. The second algorithm we present eliminates all the Normals. The algorithms we employ are the standard algorithms in non-self referential reasoning adapted to our self-referential questioning. In fact both cases can be treated simultaneously using the abstract approach that we adopt.

Our self-referential questioning technique depends on a general principle, the "Nelson Goodman Principle," that elicits valid information from either a Knight or a Knave and we discuss it in the next section.

3.2 Nelson Goodman Principle (NGP)

Suppose we meet an individual who is either a Knight or a Knave, and we want to find out if statement P is true or false, by asking only one "Yes or No" question. Raymond Smullyan has pointed out (see Smullyan 2009, page 5) that it is indeed possible to do this and he attributes this observation to the logician and philosopher, Nelson Goodman, naming it the *Nelson Goodman Principle*. (We shall use the abreviation, NGP, for this principle.) Here is the question to be asked to either a Knight or a Knave:

(Q) "Is it the case that P if and only if you are a Knight?"

It is easy to see that a "Yes" answer to question Q from either a Knight or a Knave means that P is true; for, if a Knight answers "Yes," Q must be true; also, since "you are a Knight" is true, P must be true as well. If a Knave answers "Yes," Q must be false; since "you are a Knight" is false, P cannot be false, since then the biconditional would be true! Hence P is true in this case as well.

On the other hand, a "No" answer to Q from a Knight or Knave means that P is false; for, if a Knight answers "No," Q must be false and since "you are a Knight" is true, P must be false. If a Knave answers "No," Q must be true, and since "you are a Knight" is false, P must also be false.

3.3 A Unified Approach

In both the standard treatments and our self-referential treatment, the set of individuals is divided into "reliable" individuals, $\boldsymbol{R}$, and "unreliable" individuals, $\boldsymbol{U}$. In the standard versions, the reliables always tell the truth and the unrealiables don't. In our self-referential version, both those that always tell the truth and those that always lie are considered "reliable;" all others are considered "unreliable." In both cases, "Yes or No" questions, Q(X,Y) are constructed and asked of individuals, X, about individuals, Y. In the usual case, without self-referential reasoning, Q(X,Y) is often : " X, is Y reliable?" In the self-referential case, Q(X,Y) might be: "X, is Y reliable if and only if you are a Knight?" It should be noted that in both cases, (i) a "Yes" answer implies that either X is in $\boldsymbol{U}$ or Y is in $\boldsymbol{R}$; while, (ii) a "No" answer implies X is in $\boldsymbol{U}$ or Y is in $\boldsymbol{U}$. In fact it is just these simple properties that are used in proofs that mimimize the number of "Yes or No" questions that need to be asked. Dually, we sometimes want: Q(X, Y): "X, is Y unreliable?"; or, in the self-referential case: "X, is Y unreliable if and only if you are a Knight?" In these two cases, (i) a "Yes" answer implies X is in $\boldsymbol{U}$ or Y is in $\boldsymbol{U}$, while, (ii) a "No" answer implies X is in $\boldsymbol{U}$ or Y is in $\boldsymbol{R}$. (I am indebted to Mark Wildon for suggesting the adoption of this abstract approach after reading an earlier draft of this paper.)

Using these observations, we can give generalized proofs that treat the ordinary and self-referential cases simultaneously. Our proofs follow standard proofs from the literature quite closely. To find one reliable individual, we adapt the lovely proof given by Taschuk (2012). To identify all the individuals we follow the proof sketched by Blecher (1983).

3.4 Finding a Reliable Individual

Theorem 1 *Let* $\boldsymbol{I} = \boldsymbol{R} \cup \boldsymbol{U}$, *with* $\boldsymbol{R} \cap \boldsymbol{U} = \emptyset$. *Let Q(X,Y) be a "Yes or No" question such that if X is asked this question about Y, (a) a "Yes" answer implies that either*

$X \in \boldsymbol{U}$ or $Y \in \boldsymbol{R}$ and (b) a "No" answer implies that $X \in \boldsymbol{U}$ or $Y \in \boldsymbol{U}$. Suppose that $|\boldsymbol{U}| < |\boldsymbol{R}|$. Then an element of $\boldsymbol{R}$ can be found by asking Q(X,Y) at most $n - b(n)$ times, where $n = |\boldsymbol{I}|$ and $b(n)$ is the number of 1s in the binary representation of n.

Proof (1) Arbitrarily group the elements of I into ordered pairs with possibly one left over. (2) For each pair (X,Y) ask the first person, X, the question, Q(X,Y). If the answer is "No," remove both members of the pair from further consideration. If the answer is "Yes," remove just the first element. (3) If there was an unpaired element, either keep or remove this element, so as to maintain an odd number of elements. We claim that after completing these three steps, it is still the case that fewer than half of the remaining elements of $\boldsymbol{I}$ are in $\boldsymbol{U}$.

Proof of Claim. Let r be the number of elements in $\boldsymbol{R}$, initially and r_2, the number of elements in $\boldsymbol{R}$ after step (2), excluding the unpaired element, if any, and let r_3 be the number of elements in $\boldsymbol{R}$ after step (3). Similarly we define u, u_2, u_3, as the numbers of elements of $\boldsymbol{U}$, initially and after steps (2), (3), respectively. Then $r > u$ and we wish to show that $r_3 > u_3$.

Let $\#(\boldsymbol{R},\boldsymbol{R})$ be the number of ordered pairs produced by step (1) with both members in $\boldsymbol{R}$; let $\#(\boldsymbol{U},\boldsymbol{U})$ be the number of pairs with both members in $\boldsymbol{U}$; let $\#(\boldsymbol{R},\boldsymbol{U})$ be the number of ordered pairs whose first member is in $\boldsymbol{R}$ and whose second member is in $\boldsymbol{U}$; let $\#(\boldsymbol{U},\boldsymbol{R})$ be the number of ordered pairs whose first member is in $\boldsymbol{U}$ and whose second member is in $\boldsymbol{R}$.

In step (2), every (R,R) pair, $R \in \boldsymbol{R}$, must answer "Yes" (since a "No" answer would imply $R \in \boldsymbol{U}$!); so the second element of $\boldsymbol{R}$ remains; this implies $r_2 \geq \#(\boldsymbol{R},\boldsymbol{R})$. However, the only way for an element of $\boldsymbol{U}$ to survive step (2) is to be the second element of a $(\boldsymbol{U},\boldsymbol{U})$ pair with the first element answering "Yes." This implies that $u_2 \leqslant \#(\boldsymbol{U},\boldsymbol{U})$. There are three cases to be considered.

Case 1. $\boldsymbol{I}$ is even. Then $r = 2\,\#(\boldsymbol{R},\boldsymbol{R}) + \#(\boldsymbol{R},\boldsymbol{U}) + \#(\boldsymbol{U},\boldsymbol{R})$ and $u = 2\,\#(\boldsymbol{U},\boldsymbol{U}) + \#(\boldsymbol{R},\boldsymbol{U}) + \#(\boldsymbol{U},\boldsymbol{R})$. Therefore, since $r > u$, $\#(\boldsymbol{R},\boldsymbol{R}) > \#(\boldsymbol{U},\boldsymbol{U})$. Thus, $r_2 \geq \#(\boldsymbol{R},\boldsymbol{R}) > \#(\boldsymbol{U},\boldsymbol{U}) \geq u_2$, that is, $r_2 > u_2$. There is no unpaired element, so $r_3 > u_3$.

Case 2. $\boldsymbol{I}$ is odd and the unpaired element belongs to $\boldsymbol{R}$. Then, $r = 2\,\#(\boldsymbol{R},\boldsymbol{R}) + \#(\boldsymbol{R},\boldsymbol{U}) + \#(\boldsymbol{U},\boldsymbol{R}) + 1$ and $u = 2\,\#(\boldsymbol{U},\boldsymbol{U}) + \#(\boldsymbol{R},\boldsymbol{U}) + \#(\boldsymbol{U},\boldsymbol{R})$. Hence, since $r > u$, $\#(\boldsymbol{R},\boldsymbol{R}) \geq \#(\boldsymbol{U},\boldsymbol{U})$ and this, together with $r_2 \geq \#(\boldsymbol{R},\boldsymbol{R})$ and $u_2 \leqslant \#(\boldsymbol{U},\boldsymbol{U})$, implies $r_2 \geq u_2$. If $r_2 > u_2$, then $r_3 > u_3$, whether or not we keep the unpaired element of $\boldsymbol{R}$. If $r_2 = u_2$, then the total number of elements of $\boldsymbol{I}$ kept after step (2) is even and so we keep the unpaired element of $\boldsymbol{R}$ and $r_3 = r_2 + 1 > u_2 = u_3$.

Case 3. $\boldsymbol{I}$ is odd but the unpaired element was in $\boldsymbol{U}$. Then, $r = 2\,\#(\boldsymbol{R},\boldsymbol{R}) + \#(\boldsymbol{R},\boldsymbol{U}) + \#(\boldsymbol{U},\boldsymbol{R})$ and $u = 2\,\#(\boldsymbol{U},\boldsymbol{U}) + \#(\boldsymbol{R},\boldsymbol{U}) + \#(\boldsymbol{U},\boldsymbol{R}) + 1$. Therefore, since $r > u$, $2\,\#(\boldsymbol{R},\boldsymbol{R}) > 2\,\#(\boldsymbol{U},\boldsymbol{U}) + 1$. So $\#(\boldsymbol{R},\boldsymbol{R}) > \#(\boldsymbol{U},\boldsymbol{U})$. Thus, $r_2 \geq \#(\boldsymbol{R},\boldsymbol{R}) > \#(\boldsymbol{U},\boldsymbol{U}) \geq u_2$, that is, $r_2 > u_2$. If $r_2 = u_2 + 1$, then the total number of elements of $\boldsymbol{I}$ kept after step (2) is odd and so the unpaired element of $\boldsymbol{U}$ is removed; so $r_3 = r_2$, $u_3 = u_2$ and $r_3 > u_3$. If $r_2 > u_2 + 1$, then $r_3 > u_3$, no matter what.

Each time this three step procedure is repeated, at least half of the elements are removed from consideration. Finally, the sole remaining element must be in $\boldsymbol{R}$.

Surely, more questions will need to be asked, when in step (2) only one person is removed from each pair because all questioned answer "Yes." This implies that the number of questions asked is bounded by the number of "Yes" answers received to Q(X,Y) in step (2). We claim that this upper bound to the number of questions is $n - b(n)$, where $b(n)$ is the number of 1s in the binary representation of n. Assume that this is true for all $k < n$. If n is even, step (1) results in $n/2$ pairs with no one left over, $n/2$ questions having already been asked and $n/2$ remain (assuming all "Yes" answers). Since $n/2 < n$, our induction assumption implies that the number of further questions that are needed is bounded by $n/2 - b(n/2)$. However the binary representation of $n/2$ has same number of 1s as the binary representation of n. Thus the total number of questions asked is at most $n/2 + n/2 - b(n/2) = n - b(n)$. If n is odd, $b(n) = b(n\text{-}1) + 1$. Step (1) results in $\lfloor n/2 \rfloor$ (the floor of $n/2$) pairs and $b(\lfloor n/2 \rfloor) = b(n\text{-}1)$. Thus the total number of questions, when n is odd, is bounded by $\lfloor n/2 \rfloor + \lfloor n/2 \rfloor - b(\lfloor n/2 \rfloor) = (n - 1) - b(n\text{-}1) \leq n - b(n)$. Thus, the total number of questions asked is bounded by $n - b(n)$.

Corollary. Suppose that there are n processors, u of which are unreliable, where $u < n/2$. Then at most $n - b(n)$ "Yes or No" questions suffice to find a reliable processor, where $b(n)$ is the number of 1s in the binary representation of n.

Proof If only processors that are Knights are considered "reliable," let Q(X,Y) be the question: "Processor X, Is processor Y a Knight?" If both Knights and Knaves are considered reliable, use the self-referential question: "Processor X, Is Processor Y reliable if and only if you are a Knight?"

As has been often noted, a closely related problem is finding a majority element, by pairwise comparisons ("same" or "different"), in a set divided into two subsets– say a set of colored balls that are colored either red or blue (see Alonso et al. 1993; Saks and Werman 1991). The above proof of S. Taschuk can be modified to yield the standard results here, as well. The main idea is, in step (2), to remove both elements from the (red,blue) and (blue,red) pairs, and only the first element elements from the (red,red) and (blue,blue) pairs, etc.

3.5 Finding All the Reliables

We now turn to the problem of finding all the Reliables. Our algorithm is an adaptation of that of Blecher (1983) (rediscovered by Wildon 2010).

Theorem 2 *Let* $\boldsymbol{I} = \boldsymbol{R} \cup \boldsymbol{U}$, *with* $\boldsymbol{R} \cap \boldsymbol{U} = \emptyset$. *Let Q(X,Y) be a "Yes or No" question such that if X is asked this question about Y, (a) a "Yes" answer implies that* $X \in \boldsymbol{U}$ *or* $Y \in \boldsymbol{U}$ *and (b) a "No" answer implies that* $X \in \boldsymbol{U}$ *or* $Y \in \boldsymbol{R}$. *Let* $n = |\boldsymbol{I}|$, $u = |\boldsymbol{U}|$ *and suppose that* $u \leqslant t < n/2$. *Then all elements of* $\boldsymbol{R}$ *can be identified by asking* *the question Q(X,Y) at most* $n + t$ *times. In particular,* $(3/2)n$ *"Yes or No" questions will aways suffice if* $u < n/2$.

Proof Assume that the Theorem is true for all sets of size less than n. Choose an element Y of ***I***, whose status is unknown and ask the other elements, X, in turn, question Q(X,Y) about Y. The questioning stops as soon as either of the following two conditions is satisfied.

(a) t elements have answered "No."
(b) More elements have answered "Yes" than have answered "No."

(One of these two cases must occur, since $t < n/2$; for, if fewer than t have answered "No," then more than half answer "Yes" and so, at some point, more elements will have answered "Yes" than "No.")

Suppose the questioning stops in case (a). Then $Y \in \boldsymbol{R}$; for, if $Y \in \boldsymbol{U}$, at least one of those t, say X, that answered "No" to Q(X,Y) cannot be in ***U***, since $u \leqslant t$ and $Y \in \boldsymbol{U}$; but $X \notin \boldsymbol{U}$ implies $Y \in \boldsymbol{R}$, since X answered "No." Note that, in this case, if $t + a$ questions have been asked, with a having anwered "Yes," these a who answered "Yes" must all be elements of ***U***, since $Y \notin \boldsymbol{U}$. We then ask processor Y about the other $(n - a)$ elements, Z, using question Q(Y,Z). A "No" answer implies $Z \in \boldsymbol{R}$; a "Yes" answer implies $Z \in \boldsymbol{U}$. In this way we have identified all the reliable elements of ***I*** using $(n - a) + (t + a) = n + t$ questions.

Suppose the questioning stops in case (b) and ***C*** is the set of those elements X in ***I*** who have answered "Yes" to question Q(X,Y) and ***D*** is the set of those X in ***I*** who have answered "No" to Q(X,Y); thus, $|\boldsymbol{C}| = |\boldsymbol{D}| + 1$. Assume first that at least one X in ***C*** belongs to ***R***; then, the "Yes" answer to Q(X,Y) from this X implies that $Y \in \boldsymbol{U}$; moreover, all those in ***D***, having answered "No," must also be in ***U***, since $Y \notin \boldsymbol{R}$. Thus, in this case, at least $|\boldsymbol{D}| + 1$ of the elements in $\boldsymbol{C} \cup \boldsymbol{D} \cup \{Y\}$ are in ***U***, that is, at least half of the elements in $\boldsymbol{C} \cup \boldsymbol{D} \cup \{Y\}$ are in ***U***. If, however, no element of ***C*** is in ***R***, then ***C*** consists entirely of elements of ***U*** and again, at least half the elements in $\boldsymbol{C} \cup \boldsymbol{D} \cup \{Y\}$ belong to ***U***. Let $c = |\boldsymbol{C}|$. Then $|\boldsymbol{C} \cup \boldsymbol{D} \cup \{Y\}| = 2c$. Therefore, if the elements in $\boldsymbol{C} \cup \boldsymbol{D} \cup \{Y\}$ are removed from ***I***, the reduced set of elements, $\boldsymbol{E} = \boldsymbol{I} - (\boldsymbol{C} \cup \boldsymbol{D} \cup \{Y\})$, obeys the condition $u_1 \leq t - c < n_1/2$, where $n_1 = |\boldsymbol{E}|$ and $u_1 = |\boldsymbol{U} \cap \boldsymbol{E}|$. Note that $n_1 < n$. By our induction assumption, at most $n_1 + (t - c)$ questions suffice to identify all the reliable elements of ***E***. It remains to identify the reliable elements of $\boldsymbol{C} \cup \boldsymbol{D} \cup \{Y\}$. Since $(t - c) < n_1/2$, ***E*** must contain reliable elements of ***R***; let $Z \in \boldsymbol{E} \cap \boldsymbol{R}$ and ask Z the question Q(Z,Y). This uses one additional question. There are two cases.

(1) If Z answers "Yes," then $Y \in \boldsymbol{U}$. Since $Y \in \boldsymbol{U}$, those in set ***D***, that is, those X who said "No" to Q(X,Y) must be in ***U***. We then ask Z question Q(Z,X) about each element, X, in ***C***. A "Yes" answer implies $X \in \boldsymbol{U}$; a "No" answer implies $X \in \boldsymbol{R}$. This uses $c = |\boldsymbol{C}|$ additional questions.
(2) If Z answers "No," then $Y \in \boldsymbol{R}$. Since $Y \in \boldsymbol{R}$, those in set ***C***, having said "Yes" to Q(X,Y) must be in ***U***. In this case, we ask processor Z question Q about each processor X in ***D***. Again, a "Yes" answer implies $X \in \boldsymbol{U}$; a "No" answer implies $X \in \boldsymbol{R}$. This uses $c - 1$ additional questions.

Therefore, in either case (1) or case (2), we require, at most, an additional c questions. Also, before stopping in case (b), $|\boldsymbol{C}| + |\boldsymbol{D}| = n - n_1 - 1$ questions were asked. Since case (1) and case (2) are mutually exclusive, we require at most $(n - n_1 - 1) + (n_1 + t - c) + c + 1 = n + t$ questions as claimed.

Corollary 1 (Ordinary reasoning). *Suppose that there are n processors, at most t of which are Knaves and Normals, where t < n/2. Then at most n + t "Yes or No" questions suffice to find all the Knights. In particular* $(3/2)n$ *"Yes or No" questions suffice.*

Proof Let Q(X,Y) be the question: Processor X, is processor Y unreliable (a Knave or a Normal)?

Corollary 2 (Self-referential reasoning). *Suppose that there are n processors, at most t of which are Normal, where t < n/2. Then at most n + t "Yes or No" self-referential questions suffice to find all the Reliable processors (Knights and Knaves).*

Proof Let Q(X,Y) be the question: Processor X, is processor Y unreliable (a Normal) if and only if you are a Knight?

3.6 Is Anyone Really "Normal"

A former student of mine, Edward Szymkowiak, after having read an earlier version of this paper, modified the NGP to also reference the question being answered. His version, which we call "ESP" is as follows.

> (ESP) " Is it true that when you answer this question you will be telling the truth if and only if P?"

Call a Normal who decides in advance whether to lie or tell the truth *decisive*. Decisive Normals will behave like a Knight or Knave when confronted with an ESP question and hence will answer "Yes" if and only if P is true. (Of course Normals that answer randomly will not be of any use.) Hence if we know there are decisive Normals they can also be considered "Reliable."

It has been objected that it is improper to refer to the question itself within itself; however calling someone a "Knight" inside a question implies that this person will answer this same question truthfully, so I see little difference in this and ESP. Also, for computer processors, calling one a Knight seems rather extreme, since it says basically that the processor will never fail. How can we or other processors be certain of that? In fact the opposite is probably the case, that is, all processors will fail eventually. So it seems to us that using statements like ESP, which only reference the short term behavior of the processors is the safer alternative.

References

Alonso, L., Chassaing, P., Reingold, E., & Schott, R. (2004). The worst-case chip problem. *Information Processing Letters*, *89*, 303–308.
Alonso, L., Riengold, E., & Scott, R. (1993). Determining the majority. *Information Processing Letters*, *47*(5), 253–255.
Blecher, P. M. (1983). On a logical problem. *Discrete Math*, *43*, 107–110.
Hakimi, S. L., & Nakajima, K. (1984). On adaptive system diagnosis. *IEEE Transactions on Computers*, *33*, 234–240.
Hakimi, S. L., & Shmeichel, E. F. (1984). An adaptive algorithim for system level diagnosis. *Journal of Algorithms*, *5*, 526–530.
Saks, M. E., & Werman, M. (1991). On computing majority by comparisons. *Combinatorica*, *11*, 383–387.
Smullyan, R. (1978). *What is the Name of this Book?*. Englewood Cliffs, NJ: Prentice Hall.
Smullyan, R. (1992). *Satan, cantor and infinity*. Knopf Inc, New York, NY: Alfred A.
Smullyan, R. (2009). *Logical Labyrinths*. Peters, Wellsley, MA: A.K.
Taschuk, S. (2012). http://math.stackexchange.com/questions/115823/
Wildon, M. (2010). Knights, spies, games and ballot sequences. *Discrete Mathematics*, *310*, 2974–2983.

Chapter 4
Russell's Paradox, Gödel's Theorem

Melvin Fitting

Abstract Smullyan has closely examined Gödel's proof of arithmetic incompleteness both abstractly, for a mathematically sophisticated audience, and via the medium of puzzles, for a general audience. In this essentially pedagogical paper I take an intermediate position, and show that Gödel's ideas fit naturally into the context of the elementary set theory that any mathematics student would know near the beginning of studies. Gödel once wrote, about his incompleteness proof, "The analogy of this argument with the Richard antinomy leaps to the eye. It is closely related to the 'Liar' too." And further, "Any epistemological antinomy could be used for a similar proof of the existence of undecidable propositions." Here, combining the spirit of both Smullyan's formal and his popular work, I show how Russell's paradox can be used in a direct and natural way to obtain Gödel's incompleteness results.

4.1 Introduction

Self-reference is a common part of everyday language. For instance this sentence is self-referential, but would raise no eyebrows. Then some insidious person whispers in your ear, "This sentence is not true," and your mind boggles. In its way, "This sentence is true" is even more perplexing. Yet self-reference can be a powerful tool. It leads to mathematical results, to metamathematical results, to recursive algorithms, and to paradoxes. Many logicians have an obsession with self-reference, and they would be the first to tell you so. Raymond Smullyan is in this category. His doctoral thesis, from 1959 and published as Smullyan (1961), says much about self-reference as it appears in Gödel's Theorems, Tarski's Theorem, and a family of related results. It does so very elegantly, abstracting the ideas to their bare essence. Indeed, Smullyan's work did not begin here—there are even earlier results from the late 1950's, Smullyan (1956a, b, 1957, 1958). After his dissertation he returned to the subject over and over. Sometimes his approach was formal and aimed at mathematicians, as in Smullyan

M. Fitting (✉)
Department of Computer Science, Graduate Center, City University of New York,
365 Fifth Avenue, New York, NY 10016, USA
e-mail: melvin.fitting@lehman.cuny.edu

M. Fitting and B. Rayman (eds.), *Raymond Smullyan on Self Reference*,
Outstanding Contributions to Logic 14, https://doi.org/10.1007/978-3-319-68732-2_4

(1992, 1994). Sometimes, quite remarkably, it was aimed at a non-mathematical audience through the medium of puzzles, Smullyan (1985, 1987). This is not meant to be a scholarly bibliography—our list of Smullyan's work on self-reference is far from complete.

In this paper I do not intend to analyze Smullyan's contributions, but rather to move in a somewhat different pedagogical direction. In some of his works Smullyan has reduced diagonal arguments involving self-referentiality to what is probably an abstract minimum. This certainly gets at the essence of the matter, and is vitally important for deeper understanding. But it is not particularly helpful to someone encountering the ideas for the first time. Smullyan's puzzle books are intensely mathematical without appearing so, and are marvelous at presenting diagonal arguments to people who do not realize they are doing mathematics. But what I want to do here is show how Gödel's basic ideas are intuitive and simple, given as background information that which most people with some, but not much, mathematical training would naturally have. I will follow the approach in Fitting (2007), so I will carry out the work in the framework of set theory using Russell's paradox as an organizing principle. Set theory because it was designed to be a 'universal' theory. Russell's paradox because it is something that anyone with exposure to even minimal set theory would know about. But perhaps this is already giving too much away. It is best to approach things more gradually.

4.2 Gödel Numbering and Everyday Life

Before Gödel could bring down the hammer of self-reference on Hilbert's program, he needed reference. He worked in arithmetic (more properly, a version of set theory in which he concentrated on the arithmetic part), and so he needed numbers to somehow refer to facts about numbers, where facts were represented by logical formulas. To do this he introduced what has become known as *Gödel numbering*. Numbers are thought of as codes for sequences, and formulas, and functions, and so on. Further, common operations on the things that numbers code were shown to correspond to computable functions on the numbers encoding them. This may have seemed peculiar at the time—I wasn't there and so have no direct opinion—but today it is standard fare where computers are involved. Smullyan has certainly devoted much effort to producing Gödel numberings that are simpler conceptually or computationally, and so have many other people. In Smullyan's case, *elegant* is an often-used word to describe what his work is like, and it applies here. In his most abstract treatments, coding remains behind the scenes entirely. Still, Gödel numbering takes getting used to, and is often a peculiarity at first exposure. In this section I want to discuss a little of the common human experience, and why numerical coding is actually very familiar to us.

Suppose you write down "923" and ask anyone what that is. They will say something like, "It is the number nine hundred and twenty three." Had the new math succeeded back in the 1960s perhaps they would say, "It is a numeral designating the number nine hundred and twenty three." But new math did not succeed. We find such circumlocution unnatural, although it does express what is going on more correctly. What we wrote down, "923," is not a number, it is a string that names a number—it is a code for a number. We do not see numbers—they are abstract objects. We work with them through such codes. Our algorithms for adding, multiplying, and so on, all work on strings, not on numbers. Mentally, we move back and forth all the time. We say things like "923 is a three-digit number," and do not get into difficulties as a result even though it is names for numbers that have digits and not numbers themselves. Of course one could take this further and note that what we wrote down, "923," is not a string, it is what is called a "string token." If it were a string, and we wrote "923" down again, we would have two different strings, but that's not how we think. Instead we simply think we wrote the same string down twice. Like numbers, strings are abstract objects, and what we write down is intended to convey information about what abstract string we want to discuss. One can carry this too far. Ordinary language acts as if the distinction between strings and string tokens didn't exist, unless we happen to need it. So we talk about "923" as if it were a string token, a string, or a number, and don't think much about it.

What happens when we count? Suppose we have a pile of stuff and we go through it, counting as we go—in effect assigning a number to each item. Let's say we discover we have 923 things. This has nothing to do with strings. The size of the pile of stuff, the set, is 923. We can see this as a kind of coding too. The set of stuff is a code for the number 923, which also has the string "923" as its code, but in a different coding system. Things are what they are, but also we use them as codes for something else. We do so automatically and routinely. We do not think about it (except for occasions like this one). Conversation and other everyday activities would be impossible otherwise.

What is peculiar about Gödel numbering, then, is essentially psychological. Instead of using various kinds of codes for numbers, it uses numbers as codes for other things. We are somewhat accustomed to this; a referee pointed out everyday examples that most people generally are familiar with. Room numbers might use an arrangement of digits so that floor information is included. Telephone numbers commonly allow the easy read off of country and city, as well as local telephone number. These are straightforward examples. If you are a computer scientist you pass back and forth between numbers and things they code without even thinking about it. What Gödel did was create a sophisticated way in which numbers could encode other things, primarily strings. The computational aspect of Gödel numbers is central, just as our addition and multiplication algorithms operating on place-value names for numbers is central, but is completely missing in our examples involving room and telephone numbers.

As was noted above, issues that people may have with Gödel numbering at first exposure are essentially psychological. Numbers are being used for a purpose they were not 'designed' for, and this is not what we are familiar with. Unlike arithmetic, there are areas of mathematics that were specifically developed with the idea of universality, primarily set theory and category theory. One can argue that all known mathematics can be carried out using the machinery of either. One can further argue that things are more natural, or simpler, using one or another of these approaches, and people do argue about this. I will use set theory here, because people seem to be built from infancy as set theorists or category theorists, and I am not a category theorist. And indeed, we don't need much of set theory—no deep results or theorems. We basically need that set theory is intended to be universal. To quote Zermelo (1908), "Set theory is that branch of mathematics whose task is to investigate mathematically the fundamental notions 'number', 'order', and 'function', taking them in their pristine, simple form, and to develop thereby the logical foundations of all of arithmetic and analysis; thus it constitutes an indispensable component of the science of mathematics." We will see that there are sets that can be identified with numbers—that code them if you like. There are sets that can be identified with relations, or with functions, or with strings. This is common and familiar mathematical practice. We embrace it happily, and thus avoid the details of Gödel numbering.

4.3 Really Basic Set Theory

I think I can assume most readers of this paper will already know at least some set theory, at least basic notation and concepts. I believe I can safely omit an elementary tutorial. What we need is really very little. We will work with *pure* sets, that is, we will not talk about sets of chairs and tables. Moreover, all our sets will be finite. These are the most concrete of all the sets. Issues with the axiom of choice are not relevant—it is simply true for finite sets. The continuum hypothesis cannot be properly stated when only such sets are involved. Our domain is part of set theory everyone thinks they understand.

To be a bit more precise, the *hereditarily finite* sets are those sets that are finite, with finite sets as members, those having finite sets as members, and so on. The empty set is such a thing, $\emptyset$. So is $\{\{\emptyset\}, \emptyset\}$. We use $\mathbb{HF}$ for the collection of hereditarily finite sets. Informally, these are the sets that can be named using a symbol for the empty set and familiar bracket notation. We note that hereditarily finite sets have been used in an automated deduction approach to Gödel's theorems, in Paulson (2014), and have been formalized axiomatically, in Świerczowski (2003).

Many college level mathematics textbooks define relations as sets of ordered pairs, and then go on to define functions as single-valued relations. In $\mathbb{HF}$ we have a rich collection of sets. What we need in order to go further is ordered pairs. In the famous work of Russell and Whitehead on the logicist foundations of mathematics, *Principia Mathematica* Whitehead and Russell (1910), the logic of properties and the logic of properties of pairs (relations) were developed independently. Not long after publication Norbert Wiener discovered how to "define" ordered pairs, making this repeated labor unnecessary. His idea is technical, looks artificial, and simplifies things enormously. Since then other definitions have been given. Today the standard one is due to Kuratowski, and it is the one we will use. But first we should explain why we employed quotation marks a few sentences back.

Ordered pairs are a bit like numbers—we think we know what they are. We don't. What we know is what to do with them. This is something probably best captured in a category theoretic approach, but here we are using set theory. The key thing about an ordered pair is that it has a first component, a second component, these are unique, and the order matters between first and second. It is common to write $\langle x, y\rangle$ for the ordered pair with first component x and second component y. Then the fundamental fact concerning them can be summed up as follows.

$$\langle x, y\rangle = \langle a, b\rangle \text{ if and only if } x = a \text{ and } y = b \tag{4.1}$$

The notation makes it clear which is the first and which is the second component. Now, this is an operational characterization. If there is some set construction that meets (4.1) we can use it and get on with things. Kuratowski's definition is the following.

$$\langle x, y\rangle \text{ is defined to be } \{\{x\}, \{x, y\}\} \tag{4.2}$$

This makes an ordered pair into a kind of set, which is what we basically need. And it has the fundamental property (4.1), whose proof we leave to you. Hint: consider three cases: $x = y$; $a = b$; and $x \neq y$ and $a \neq b$. (You can find a proof in Fitting (2007), which is what this paper is based on. In general, that is where to look for more details about anything discussed here.)

We can think of (4.2) as giving us a *coding* of the abstract notion of ordered pair, but set theorists simply refer to $\{\{x\}, \{x, y\}\}$ as *being* an ordered pair. It is one more example of the simplifying tendencies of ordinary language—don't emphasize something if it doesn't matter. Now that we have a version of ordered pair, we can define ordered triples as $\langle x, y, z\rangle = \langle\langle x, y\rangle, z\rangle$, then ordered quadruples, and so on. They all work as we want, and have the key property that they keep us in the realm $\mathbb{HF}$.

The next thing we need is numbers, by which we mean non-negative integers. As with ordered pairs, we tend to think we know what they are, when what we really know is how they behave. And as with ordered pairs, we can find sets that will serve as convenient representatives for numbers—quite natural ones, in fact. We want the number 3, say, to be a set with 3 elements, so it can be used as a measure of size, and we want to do this in a simple way, for all numbers. Von Neumann, in (1923), introduced a simple way of representing the *ordinal* numbers, finite and infinite which is more than we need, and we use his idea here. Just let each number be the set of all smaller numbers. There are no numbers smaller than 0, so we take 0 to be $\emptyset$. Then 1 is $\{0\} = \{\emptyset\}$, 2 is $\{0, 1\} = \{\emptyset, \{\emptyset\}\}$, and so on. The structure gets complicated as we go on, but underneath, here's what is going on. Suppose we have defined a set to act as the number n; let us just say *to be* the number n. Then we want $n + 1$ to be the set of numbers smaller than $n + 1$, that is, the set of numbers smaller than or equal to n. We already have the set of numbers smaller than n; it is what we are calling n. So what we want is just $n \cup \{n\}$. Putting all this together, the numbers are the smallest subset $\mathbb{N}$ of $\mathbb{HF}$ such that:

$$\emptyset \in \mathbb{N}$$
$$\text{if } n \in \mathbb{N} \text{ then } n \cup \{n\} \in \mathbb{N}$$

We call $x \cup \{x\}$ the *successor* of x. Of course the sets just defined are not numbers, they are sets. But since numbers are abstract, and these are convenient set representatives of them, set theorists simply refer to them as numbers, and so shall we. We introduce the usual symbols 0 for $\emptyset$, 1 for the successor of 0, 2 for the successor of 1, and so on as long as useful to us.

Notice that each number, as just characterized, is a member of $\mathbb{HF}$—it is a hereditarily finite set. The set of *all* numbers is infinite, and is not a member of $\mathbb{HF}$. It is, however, a subset, and a fairly well-behaved one at that. If number x is less than number y (in the everyday sense), we can say the ordered pair $\langle x, y\rangle$ is in the less-than relation. This ordered pair is a member of $\mathbb{HF}$. The set of all such ordered pairs, which we can identify with the less-than relation itself, is not a member. It is, however, an infinite subset of $\mathbb{HF} \times \mathbb{HF}$, the collection of all ordered pairs with components from $\mathbb{HF}$. We will say it is a relation on $\mathbb{HF}$. What we will be examining next is subsets of $\mathbb{HF}$, and relations and functions on it.

4.4 The Machinery of Logic

We have been talking about special subsets of $\mathbb{HF}$, like numbers, and relations on $\mathbb{HF}$, like less-than. But we have been doing so in natural language, with all its warm fuzziness. In order to make the discussion more precise we need the machinery of formal logic. Just as I did with set theory, I will assume most readers of this paper already have some acquaintance with logic—not much is needed for our purposes. We will use (first-order) formulas, with terms that involve constant symbols, function symbols, and variables. Atomic formulas are relation symbols applied to the appropriate number of terms. Then formulas are built up using logical connectives and quantifiers. We will use $\wedge$ (and), $\vee$ (or), $\neg$ (not), and $\supset$ (material implication) as logical connectives, and $\forall$ and $\exists$ as universal and existential quantifiers. An informal understanding of free and bound variable occurrence will be enough for our purposes.

We want a formal language appropriate for discussing $\mathbb{HF}$. Bracket notation such as $\{a, b, c\}$, handy as it is in natural language, is not something we can use. Essentially, the notation represents a function symbol, mapping sets to a set containing them as members, but the number of inputs can vary. Functions with varying numbers of inputs are not a standard part of first-order logic. Fortunately there is a work-around. Consider the two-place function on $\mathbb{HF}$ defined by $\mathscr{A}(x, y) = x \cup \{y\}$. Every member of $\mathbb{HF}$ can be generated from $\emptyset$ using this function. For example, consider the set $\{1, 2\}$. We have

$$
\begin{aligned}
1 &= \{0\} = \emptyset \cup \{\emptyset\} = \mathscr{A}(\emptyset, \emptyset) \\
2 &= 1 \cup \{1\} = \mathscr{A}(1, 1) \\
\{1\} &= \emptyset \cup \{1\} = \mathscr{A}(\emptyset, 1) \\
\{1, 2\} &= \{1\} \cup \{2\} = \mathscr{A}(\{1\}, 2)
\end{aligned}
$$

and hence, unwinding all this

$$\{1, 2\} = \mathscr{A}(\mathscr{A}(\emptyset, \mathscr{A}(\emptyset, \emptyset)), \mathscr{A}(\mathscr{A}(\emptyset, \emptyset), \mathscr{A}(\emptyset, \emptyset)))$$

We didn't show you this just to scare you. The point is, it can be done, and beyond that the details don't need to concern us much.

In our formal logic language we assume we have a constant symbol intended to represent the empty set—we will simply use $\emptyset$ for both the symbol and the thing it is supposed to stand for. Context should keep things straight. Similarly we have a two place function symbol intended to represent the function $\mathscr{A}$ discussed above. We use "$\mathscr{A}$" as notation for both the symbol and what it stands for, and let context sort it out. Let us, very briefly, be a little more careful. We are in the middle of specifying a formal language. We have two pieces so far, a constant symbol and a function symbol. Logical languages can have many models. In our case one of these models is the *intended* model. This model is $\mathbb{HF}$, with the constant symbol $\emptyset$ interpreted by the member $\emptyset$ of $\mathbb{HF}$, and with the function symbol $\mathscr{A}$ interpreted by the function $\mathscr{A}$ on $\mathbb{HF}$. We overload notation to make reading more natural, trusting that it does no harm. In this paper the intended model is the only one we are ever interested in.

Now we continue with the formulation of our logical language, and its intended interpretation. We want one relation symbol, intended to represent the membership relation on $\mathbb{HF}$. We overload $\in$ for this purpose, and we write it in the familiar infix position. Thus, if t and u are terms, $(t \in u)$ is an atomic formula. This completes the basic machinery of our logical language. Now formulas, substitution for free variables, and all the rest, are exactly as in any logic textbook. Following Fitting (2007) we call this formal language LS, because it is intended to be a language for set theory.

Generally in logic one works with many models, and must interpret formal symbols in each one. Here we are only interested in one model for LS, the intended one $\mathbb{HF}$. This makes things rather simple to explain. Indeed, we only need to discuss how things work when no free variables are involved, and that makes things very simple indeed.

Each term of LS with no variables designates a member of $\mathbb{HF}$. When working with it, the symbol $\emptyset$ designates the set $\emptyset$, the function symbol $\mathscr{A}$ designates the function $\mathscr{A}$, and we just do the calculations. We gave an example of this above when we determined that the term $\mathscr{A}(\mathscr{A}(\emptyset, \mathscr{A}(\emptyset, \emptyset)), \mathscr{A}(\mathscr{A}(\emptyset, \emptyset), \mathscr{A}(\emptyset, \emptyset)))$ designated the set $\{1, 2\}$. This can all be made more precise, and we will come back to it later on. It is common to call a term with no free variables a *closed term*. Similarly a *closed formula* is a formula with no free variables. We will employ this terminology from now on. The upshot of the discussion so far is: every closed term of LS designates a member of $\mathbb{HF}$, and every member of $\mathbb{HF}$ is designated by a closed term of LS.

We should note that, except for $\emptyset$, members of $\mathbb{HF}$ are designated by many closed terms. For instance, $\mathscr{A}(\emptyset, \emptyset)$ designates $\emptyset \cup \{\emptyset\}$, or just $\{\emptyset\}$, in standard bracket notation. We leave it to you to check that $\mathscr{A}(\mathscr{A}(\emptyset, \emptyset), \emptyset)$ designates the same thing. This multiplicity of designation does no harm, and we mostly ignore it.

We now turn to the task of characterizing which closed formulas of LS are *true* over (or about or in) $\mathbb{HF}$, our standard model. The fact that all members of $\mathbb{HF}$ are designated by closed terms makes this rather easy.

Suppose $(t \in u)$ is a closed atomic formula. It is true in the standard model just in case the set in $\mathbb{HF}$ designated by t is a member of the set designated by u.

If $(X \wedge Y)$ is a closed formula, it is true in the standard model just in case both X and Y are. And similarly for the other propositional connectives.

Quantifiers are easy to handle because all members of $\mathbb{HF}$ have names in LS. Let us write $\varphi(x/t)$ to denote the result of replacing all free occurrences of variable x in φ with occurrences of t, where t is a closed term. Suppose $(\forall x)\varphi$ is a closed formula of LS. It is true in the standard model $\mathbb{HF}$ just in case $\varphi(x/t)$ is true for every closed term t. Similarly if $(\exists x)\varphi$ is a closed formula of LS, it is true in $\mathbb{HF}$ just in case $\varphi(x/t)$ is true for some closed term t.

We have now characterized which closed formulas of LS are true in $\mathbb{HF}$, and that's it for the machinery we need from logic.

4.5 What Can LS Say?

We have defined significant subsets of $\mathbb{HF}$, the set of numbers, the set of ordered pairs, and so on. Our definitions have been in the usual mathematical style consisting of a mixture of natural language and mathematical symbolism. Now we will show that much of this can be carried out more formally, using LS. Suppose $\varphi(x)$ is a formula of LS with only the variable x free. We can think of it as *representing* a subset of $\mathbb{HF}$, loosely the set of x in $\mathbb{HF}$ for which $\varphi(x)$ is true. We are going to examine which subsets of $\mathbb{HF}$ are representable by formulas in this way, but we have some comments first.

We keep the clutter down by leaving structured terms, such as $\mathscr{A}(v_0, \emptyset)$, out of the picture. We will attempt to represent sets using formulas that do not use structured terms—they will be built up from variables, but nothing more complicated. Actually this is no real restriction, as items Rep 3 and Rep 4 below will show. But we do not go as far as we might. For many purposes it is important to show not only that certain sets and relations are representable, but that they are represented by formulas that are especially simple in the way they use quantifiers. In our work here, this will not play a role and so we ignore the point. However if you happen to look at the fuller discussion in Fitting (2007), you will see much attention is paid to keeping formulas simple.

Although structured terms will not appear in representing formulas, they do play a role in how we evaluate these formulas. In our formal language we work with hereditarily finite sets via names for them, closed terms, and this simplifies things overall.

As we go on in this section we begin leaving details out, or leaving them to you. It is important to know various things are representable. Exactly how doesn't matter much. It would be good to try a few examples yourselves, though.

Representability Let $\varphi(x)$ be a formula of LS with only x free, and *with only variables allowed as terms*—we call this a *representing* formula. We say $\varphi(x)$ *represents* the subset of $\mathbb{HF}$ consisting of all sets s such that for some closed term t that designates s, $\varphi(x/t)$ is true in $\mathbb{HF}$. A subset of $\mathbb{HF}$ that has a representing formula is a *representable* subset. In a similar way representing formulas with two free variables represent two-place relations on $\mathbb{HF}$, and so on.

For representability purposes, it doesn't matter if we require $\varphi(x/t)$ be true for *some* term t designating s or for *every* such term. The two are equivalent.

Here are a few basic examples of representable sets and relations.

Rep 1 The subset relation: $\{\langle x, y\rangle \mid x \subseteq y\}$. This is represented by the following formula with two free variables: $(\forall z)(z \in x \supset z \in y)$.

Rep 2 Equality: $\{\langle x, y\rangle \mid x = y\}$. Because $x = y$ if and only if $(x \subseteq y) \wedge (y \subseteq x)$, where each conjunct (which is not actually a proper formula), is replaced with its representing formula from Rep 1.

Rep 3 Being the empty set. Because $x = \emptyset$ if and only if $\neg(\exists y)(y \in x)$.

Rep 4 The $\mathscr{A}$ function (as a three-place relation). Because $\mathscr{A}(y, z) = x$ if and only if $x = y \cup \{z\}$ if and only if $(y \subseteq x) \wedge (z \in x) \wedge (\forall w)(w \in x \supset (w \in y \vee w = z))$, where each piece of this that is not properly a formula is replaced with its representing formula as defined above.

We know every member of $\mathbb{HF}$ can be generated using $\emptyset$ and $\mathscr{A}$. Then using Rep 3 and Rep 4 above it is easy to see that, for any member $s \in \mathbb{HF}$, being s is representable. That is, there is a representing formula, and the set that it represents consists of just s. You might try a few examples. And here are a few more items you might try on your own.

Rep 5 Being transitive: x is transitive if and only if every member of x is also a subset of x.

Rep 6 Unordered pair: $x = \{y, z\}$.

Rep 7 Ordered pair: $x = \langle y, z\rangle$. Also, the property of being an ordered pair.

Rep 8 (Finite) binary relation: x a set of ordered pairs.

Rep 9 (Finite) function: x a single-valued set of ordered pairs. That is, there are no distinct ordered pairs in x having the same first component.

The set of numbers is also representable, but we cannot use the characterization of them from Sect. 4.3 which was that numbers are members of the smallest subset of $\mathbb{HF}$ containing 0 and closed under successor. While this characterization is the natural one, when analyzed it turns out to involve quantification over *subsets* of $\mathbb{HF}$, and we only have quantification over *members*. (Set quantification comes in when characterizing the notion of *smallest*.) But in $\mathbb{HF}$ we have picked a particular representation of numbers; they are sets of a very special kind. Any representation of an abstract concept will have some special attributes that other representations do not share. For instance, our common place-value way of denoting numbers using strings has the peculiarity that every number name, reversed, is another number name. Suppose we designated numbers by writing 0 followed by a sequence of 1's of the appropriate length. Such a representation would not share this reversal property. Reversability is an artifact of our representation, not something essential. Well, it turns out that our way of representing numbers as sets has an artifact that completely characterizes them, and we can make use of it now. The notion of *transitive* was mentioned above, as Rep 5. A set is transitive if every member is a subset. It turns out that the subcollection of $\mathbb{HF}$ consisting of numbers is exactly the collection of those members of $\mathbb{HF}$ that are transitive and that have only transitive members. We do not show this here—a full proof can be found in Fitting (2007).

Rep 10 The set of numbers is representable, by the formula $\mathsf{transitive}(x) \wedge (\forall y)(y \in x \supset \mathsf{transitive}(y))$, where $\mathsf{transitive}(z)$ is replaced with its representing formula, which you produced if you did Rep 5.

By the way, our particular representation of numbers was based on the idea that each number is the set of smaller numbers. Then the less-than relation on numbers coincides with the membership relation, and so it is obviously representable.

A *finite sequence* is a function whose domain is a finite initial segment of the numbers, say $\{0, 1, \ldots, n\}$. We have arranged things so that $\{0, 1, \ldots, n\}$ is just $n + 1$, so we can simplify things and say a finite sequence is a function whose domain *is* a number. In fact, the number that is the domain of a finite sequence is also the length of the sequence. If f is a finite sequence and k is in its domain, we call $f(k)$ the *kth* term of the sequence and we denote it by f_k in the usual way. We leave it to you to show the following are representable.

Rep 11 Finite sequence with length given: x is a finite sequence of length (domain) y.

Rep 12 Finite sequence: x is a finite sequence (of some length).

Rep 13 Occurs in a finite sequence at a particular location: x is a finite sequence, the sequence has an nth term, and the nth term is y. Similarly, occurs somewhere in a finite sequence.

Now things get harder, but we have finite sequences to help us. Before getting to what we are really interested in, let us look at a simple example that illustrates a general idea. Let us sketch why the set of even numbers is representable. We don't yet have multiplication, so we can't say the even numbers are those that are twice some number, but we can proceed along the following lines. The number 8 is even because it is the last term of the finite sequence 0, 2, 4, 6, 8, and this sequence is easily described as one in which the initial term is 0, and otherwise each term is 2 more than its predecessor. It is easy to see that a number x is 1 more than a number y if $x = \mathscr{A}(y, y)$. Suppose we abbreviate this by $x = y^+$ to make reading easier. Then we can say x is 2 more than y by saying $x = (y^+)^+$. But then we can represent the even numbers by saying we have a finite sequence that ends with the given number, whose first term is 0, and in which every other term x is $(y^+)^+$, where y its predecessor. Here is a formula (abbreviated, of course) that does this. (We will not use this formula after our current motivational discussion.) e is even if:

$$
\begin{aligned}
&(\exists s)(\exists n)[\\
&\qquad s \text{ is a finite sequence of length } n \wedge\\
&\qquad \text{the last term of } s \text{ is } e \wedge\\
&\qquad \text{the 0th term of s is } 0 \wedge\\
&\qquad (\forall m)((m < n \wedge \neg(m = 0)) \supset\\
&\qquad\quad (\exists k)(\exists x)(\exists y)(m = k^+ \wedge\\
&\qquad\qquad\qquad \text{the } k\text{th term of } s \text{ is } y \wedge\\
&\qquad\qquad\qquad \text{the } m\text{th term of } s \text{ is } x \wedge\\
&\qquad\qquad\qquad x = (y^+)^+))\\
&\qquad]
\end{aligned}
$$

This use of finite sequences is fundamental, and we now apply the idea to show addition is representable. One can also represent multiplication, exponentiation, and so on, but addition is what we need for Rep 15, so we bring it in now. The idea is that to add 4 and 3 say, we count off 3 numbers after 4. Counting, here, is moving from a number to its successor and as noted above, the successor of a number n is simply $\mathscr{A}(n, n)$. As to the rest of it, to count off 3 numbers after 4, we form a finite sequence of length 3, starting at 5 (the successor of 4), and in which each item from then on is formed by taking the successor of the item before it. The last term in the sequence is then $4 + 3$. We leave the formal details to you, with the evens as a model.

Rep 14 Addition: as a three place relation, x, y, and z are numbers and $x + y = z$.

The next item takes some work, but the basic ideas are straightforward given the machinery shown to be representable thus far.

Rep 15 Concatenation: x, y, and z are finite sequences, the length of z is the length of x + the length of y, and z begins with the items in sequence x followed by the items in sequence y.

It has turned out that, with formulas of LS, we can say a great deal of interest concerning $\mathbb{HF}$—important sets and relations are representable. Next it is time to discuss the formal machinery of LS itself.

4.6 Logic Is Mathematics Too

Set theory was designed to be something in which mathematics could be formalized (with the finitary parts landing in $\mathbb{HF}$). At least since David Hilbert in the early parts of the twentieth century, logic has been seen as part of mathematics, with considerable portions being finitary in nature. Now we want to sketch how the finitary parts of logic are, in fact, representable in $\mathbb{HF}$. This is the most technical section of the paper. Still, the basic ideas are not hard to grasp—it is the details that make it technical. We leave much of these details to be filled in by the reader, and encourage the effort.

A proof, as a formal object, is often taken to be a sequence of formulas meeting certain consequence conditions. Well we have finite sequences in $\mathbb{HF}$, so what we need are formulas. A formula is a finite sequence of symbols, again meeting certain conditions. So what we need are symbols. A symbol is really quite an arbitrary thing. When using computers, symbols are thought of as strings of 0's and 1's. In books, symbols are represented by particular typographic shapes. We are working with sets, so symbols should be sets of some special kind. We need to be able to pick them out, but otherwise their makeup is quite arbitrary. We take them to be certain ordered pairs, following the pattern used in Fitting (2007), a pattern which is displayed in Table 4.1. It is easy to show that the set of variables is representable, using material from the previous section. This is the only infinite category. Being a connective is also representable, which is easy to show because the connectives constitute a finite set—see the comments following Rep 4. Similarly for each of the other categories. Notice, for instance, that the *symbol* $\emptyset$ is, here, the set $\langle 6, 0\rangle$, where 0 is $\emptyset$ and 6 is built up using it. Our symbol $\emptyset$ and $\emptyset$ the empty set are obviously different things which context will (hopefully) disambiguate. Next we need to define *terms*, things built up from variables and the constant symbol $\emptyset$ using the function symbol $\mathscr{A}$. When you

Table 4.1 The symbols of LS

Variables	v_0 v_1 v_2 $\cdots$ $\langle 0, 0\rangle$ $\langle 0, 1\rangle$ $\langle 0, 2\rangle$ $\cdots$
Connectives	$\neg$ $\wedge$ $\vee$ $\supset$ $\langle 1, 0\rangle$ $\langle 1, 1\rangle$ $\langle 1, 2\rangle$ $\langle 1, 3\rangle$
Quantifiers	$\forall$ $\exists$ $\langle 2, 0\rangle$ $\langle 2, 1\rangle$
Punctuation	) (, $\langle 3, 0\rangle$ $\langle 3, 1\rangle$ $\langle 3, 2\rangle$
Relation symbol	$\in$ $\langle 4, 0\rangle$
Function symbol	$\mathscr{A}$ $\langle 5, 0\rangle$
Constant symbol	$\emptyset$ $\langle 6, 0\rangle$

read this paper you might see an assertion like "$\mathscr{A}(v_0, \emptyset)$ is a term." More properly, it is a typographic string representing a term. But in our set theoretic context this term is the finite sequence of length 6, let us call it a, where a_0 is $\mathscr{A}$, a_1 is (, a_2 is v_0, a_3 is the comma, a_4 is $\emptyset$ (the symbol), and a_5 is). This is a function with domain 6, that is, domain $\{0, \ldots, 5\}$, whose value at i is the ith symbol of $\mathscr{A}(v_0, \emptyset)$.

More generally, atomic formulas, and formulas, are similar constructs. Let us say a *string* is a finite sequence of symbols of LS. The term discussed above, $\mathscr{A}(v_0, \emptyset)$, is a string and similarly, atomic formulas and formulas are strings. We need to show each of these categories of strings is representable. We sketch the ideas, omit the details, and continue the numbering pattern introduced in Sect. 4.5.

One begins with the notion of *formation sequences for terms*, which shows how a term is built up. (This is a common construct in logic textbooks.) A formation sequence for a term is a sequence of strings in which each item is: a length 1 string consisting of just a variable, a length 1 string consisting of just the constant symbol $\emptyset$, or else is a string consisting of the $\mathscr{A}$ function symbol, followed by the (symbol, followed by an earlier item in the formation sequence, followed by the comma punctuation symbol followed by an earlier item in the formation sequence followed by the) symbol. A term, then, is simply something that appears in a formation sequence for terms. Then it is enough to show the set of formation sequences for terms is representable, and this can be done using the machinery outlined in the previous section, primarily Rep 15, though it takes some work and care. This, and all other constructs mentioned in this section, are discussed in detail in Fitting (2007). We thus have that the following are representable.

Rep 16 Being a formation sequence for terms.
Rep 17 Being a term.
Rep 18 Being a closed term. (This is handled the same way terms are, but variables are not allowed in formation sequences.)

Next we need representability of atomic formulas. These are finite sequences that consist of the symbol (, followed by a term, followed by the symbol $\in$, followed by a term, followed by the symbol). Using Rep 15 and Rep 17 is rather easy to show the set of atomic formulas is representable. We record this below. We also need, for use later on, that being an atomic formula with no structured terms, only variables, is representable. This is easier than being an atomic formula, in fact.

Rep 19 Being an atomic formula.
Rep 20 Being an atomic formula with variables but no structured terms.

We can push this significantly further. We know closed terms designate members of $\mathbb{HF}$, and we can calculate which ones. Well, we can do this formally in the following way. Suppose t is a closed term. Construct a formation sequence for t, and simultaneously construct a sequence of sets so that the nth item in the formation sequence for t designates the nth set in the set sequence. This is actually quite easy. For instance, if the nth item in the formation sequence is built using the symbol $\mathscr{A}$ and the ith and jth items in the formation sequence, both of which precede the nth, make the nth item in the set sequence the set $\mathscr{A}(a, b)$ where a is the ith and b is the

jth item in the set sequence. Similarly for $\emptyset$. In this way we get representability of the following relation.

Rep 21 x is a closed term that designates set y.

Following a common textbook approach, formulas are things that appear in *formation sequences for formulas*, which are similar to formation sequences for terms. In such a sequence every item is: an atomic formula, or is built from an earlier item using the negation symbol, or is built from two earlier items using a connective symbol, or is built from an earlier item using a quantifier symbol and a variable symbol. With some work, which is tedious but straightforward, the set of formation sequences for formulas can be shown to be representable by a formula. It then follows easily that the set of formulas itself is representable. This gives us Rep 22 and Rep 23.

One can elaborate things a bit, and keep track of free variables while constructing a formula. Here is the idea. Instead of one sequence, we construct two in parallel, call them f and g. In f we follow the rules for a formation sequence. Simultaneously in g we produce sets of variables so that g_k is the set of free variables of formula f_k. For example, one condition is that if f_k is the conjunction of f_i and f_j, then set g_k is the union of sets g_i and g_j. Another condition is that if f_k is the result of applying quantifier $(\forall x)$ to f_i then set g_k is set g_i with x removed. You can figure out the other conditions for yourself and, once you have them, writing an LS formula to capture what is going on is rather straightforward. In this way we can get Rep 24 below. Similarly we can manage things so that no structured terms appear, only variables. We do this by working with formation sequences in which atomic formulas meet condition Rep 20 and not just Rep 19.

Rep 22 Formation sequence for a formula.
Rep 23 Being a formula.
Rep 24 Being a formula with a certain set of free variables.
Rep 25 Being a formula with one (specified) free variable.
Rep 26 Being a formula with one (specified) free variable and no structured terms, that is, being a representing formula.

Using similar parallel sequence ideas, the operation of substituting a closed term for a variable in a formula can be shown to be representable over $\mathbb{HF}$. We construct a formation sequence for the original formula, and in parallel a formation sequence for the formula with the substitution carried out. We leave the details to you. This is a central notion, and you might give it some serious thought. Once you see how to do it, the following should be easy.

Rep 27 x is a formula with y as its only free variable, z is a closed term, and w is the result of replacing all free occurrences of y in x with z.

We thus have representability of all the basic syntactic machinery of formal logic.

4.7 Where Are We Now?

Self-reference is beginning to show its face, but so far it is displaying a benign aspect. Formulas of LS are, when applied to the standard model $\mathbb{HF}$, *about* sets. A formula with a free variable is perhaps true of some members of $\mathbb{HF}$, false of others, and so represents a subcollection of $\mathbb{HF}$. But a formula, as we have seen in Sect. 4.6, is also a *member* of $\mathbb{HF}$. Each formula is a set of a particular kind, and that kind is simple enough to be representable. Thus we have the curious fact that there is a formula that represents the set of formulas in $\mathbb{HF}$, and hence is true of itself!

Now let's take one more step. First-order logic, that is, logic with quantifiers over things, has a proof theory as well as a semantics. There are several equivalent ways of formulating a notion of formal proof. The simplest to describe is an axiomatic version. Certain logically valid formulas are taken to be *logical axioms*. Commonly these have a small number of forms which are easy to describe, so that the set of axioms is representable by a formula of LS. Then some *rules of inference* are adopted. Most usually, there are two. One is *modus ponens*: from X and $X \supset Y$, conclude Y. The other is *universal generalization*: From X, conclude $(\forall x)X$, where x is a variable. Both rules turn logical validities into logical validities. And both are rather easy to represent. For instance, there is a formula of LS, $M(x, y, z)$, that is true of three members of $\mathbb{HF}$ just in case x is a formula, say A, y is a formula of the form $A \supset B$, and z is B. Similarly for universal generalization.

Continuing with the characterization of axiomatic proofs, a logic proof is a sequence of formulas in which each item is either a logical axiom, or follows from two earlier items by *modus ponens*, or follows from an earlier item by universal generalization. A logical theorem is anything that appears as an item in a logical proof. Proofs, then, are something akin to formation sequences for formulas; they are sequences of formulas in which each item is justified outright or by virtue of previous items. And like formation sequences, the set of logical proofs is representable by a formula, and hence so is the set of logical theorems.

Logic alone is not enough—we often want to reason about some particular domain. In our case this is the domain $\mathbb{HF}$. Suppose we pick some set of formulas of LS to serve as non-logical axioms about $\mathbb{HF}$, call the axiom set $A_{\mathbb{HF}}$. We want to use $A_{\mathbb{HF}}$ as a basis to reason about $\mathbb{HF}$, so we allow its members to appear as items in proofs, just as we did with logical axioms above—we want to see what is provable using members of $A_{\mathbb{HF}}$ as our assumptions about sets. There is one obvious condition to be imposed: we want the members of $A_{\mathbb{HF}}$ to be true of $\mathbb{HF}$. There is another restriction that is, perhaps, not quite as obvious at first glance: we want to be able to say which formulas count as axioms. If we can't do this, we can't actually tell if something is a proof or not. We could ensure this by having only a finite set of non-logical axioms. But we could be more generous; we could specify a finite number of axiom *schemas*. An axiom scheme is a pattern, all formulas of this form count as axioms. Then we could have infinitely many axioms, with all of them falling into a finite number of patterns. This way of doing things is quite common. We can be more generous yet, and simply require that the set of axioms should be something we can describe. More formally, *the set* $A_{\mathbb{HF}}$ *of axioms should be representable.* (Requiring the set of axioms to be representable is more general than requiring a finite set of axioms, or a finite set of axiom schemes. It subsumes these as special cases.) For any candidate for a set $A_{\mathbb{HF}}$ of axioms, truth and representability seem to be the minimal conditions we could impose.

Assuming $A_{\mathbb{HF}}$ is representable, one can show the set of proofs using this set as non-logical axioms (in addition to logical axioms, *modus ponens*, and universal generalization) is representable. A proof is a sequence in which each item is: (1) a logical axiom (which is a member of a representable set whose details vary with the particular choice of axiomatization), or (2) is a member of $A_{\mathbb{HF}}$ (which is a representable set by our assumption), or (3) comes from two earlier formulas in the proof using *modus ponens* (for this we will need to make use of Rep 15 among other things), or (4) comes from an earlier item using universal generalization (we will need Rep 15 again). The details are a bit, well, detailed but if we are given a representing formula for $A_{\mathbb{HF}}$, we can write a representing formula for the set of proofs using non-logical axioms from $A_{\mathbb{HF}}$. But then we can also write a representing formula for the set of theorems provable using $A_{\mathbb{HF}}$ as axioms—a theorem is simply some formula that occurs in a proof. We thus have the following important fact.

Rep 28 The set of theorems provable axiomatically using $A_{\mathbb{HF}}$ as non-logical axioms about $\mathbb{HF}$ is a representable set provided the set $A_{\mathbb{HF}}$ itself is representable.

Of course there are notions of formal proofs other than axiomatic. There are sequent calculi, natural deduction systems, tableau systems, and so on. But all are equivalent in what they can prove, given a set of non-logical axioms, so our detailed discussion of axiomatics above is no limitation at all.

4.8 Russell's Paradox

In his famous paper Gödel (1931) Gödel observed, about his incompleteness proof for formal arithmetic, "The analogy of this argument with the Richard antinomy leaps to the eye. It is closely related to the 'Liar' too." And then in a footnote he says "Any epistemological antinomy could be used for a similar proof of the existence of undecidable propositions." In 1901, Bertrand Russell discovered his famous paradox (independently found by Ernst Zermelo in 1900, but not published). Pedagogically, Russell's argument provides us with a very natural version of the incompleteness proof, though this does not seem to have appeared in the literature except for Fitting (2007), where full details can be found. Here are the ideas.

Russell's paradox is essentially an informal argument. It is generally understood as imposing restrictions on any attempt to produce a formal version of naive set theory. Let us call a set *ordinary* if it does not belong to itself. Most everyday sets are ordinary. The set of cucumbers is not a cucumber, for instance. Now let S be the set of all ordinary sets. The paradox arises when we ask if S belongs to itself.

Any member of S must have the defining property for S, so if $S \in S$, then S must be ordinary, and then by definition, $S \notin S$. Conversely, if $S \notin S$ then S is ordinary, hence belongs to the set of ordinary sets, and so $S \in S$. Both $S \in S$ and $S \notin S$ lead to contradictions. The general conclusion is that such a set cannot exist, and set theory must be formalized in some way that prevents its existence. In Whitehead and Russell (1910), *types* were introduced for this purpose. In ZFC, the most common set theory used today, the formulation of the *Comprehension Axiom* makes it impossible to conclude that every naive property determines a set. But all this is a different story than the one that concerns us here.

Let us reformulate Russell's paradox in its starkest form. Suppose S is the set of all ordinary sets. Then, for any set x,

$$x \in S \Longleftrightarrow x \notin x \tag{4.3}$$

and since x can be any set, we can take it to be S itself. We immediately get the contradictory conclusion

$$S \in S \Longleftrightarrow S \notin S.$$

So far, this has been informal. Now let us turn formal, and shift our attention to $\mathbb{HF}$. As we have seen, some subsets of $\mathbb{HF}$ are *representable*. Recall, $R \subseteq \mathbb{HF}$ is representable if there is some representing formula $A(x)$ with one free variable such that $A(x)$ is true of exactly the members of R. More precisely, $r \in R$ if and only if for some closed term t of LS such that t designates r, $A(t)$ is true in the standard model $\mathbb{HF}$. We have seen that there are many interesting representable sets, the set of numbers and the set of formulas of LS among them. We now shift our attention from subsets of $\mathbb{HF}$ to the formulas of LS that represent them. As we saw in the previous section, formulas not only represent sets, they are sets themselves. The set of formulas itself is representable, say by $F(x)$. Then the formula $F(x)$ is actually a member of the set that it represents.

Ordinary Formulas Let $A(x)$ be a representing formula of LS, as defined in Sect. 4.5. We call the formula $A(x)$ *ordinary* if it does not belong to the set that it represents. More formally, let us introduce the convention that $\mathbf{A}$ is some closed term that designates the formula $A(x)$ (any choice of such a term will do for our purposes).

$$A(x) \text{ is ordinary} \Longleftrightarrow A(\mathbf{A}) \text{ is not true} \tag{4.4}$$

You should take a moment to observe the similarity between (4.3) and (4.4). And now, the argument that worked for Russell's paradox applies here too.

Theorem *The set of ordinary formulas is not representable by a formula.*

Proof Suppose the set of ordinary formulas were representable; say it is represented by formula $O(x)$. Then for any formula $A(x)$ with one free variable,

$$A(x) \text{ is ordinary} \Longleftrightarrow O(\mathbf{A}) \text{ is true}$$

Combining this with the definition of ordinary, (4.4), we have the following.

$$A(\mathbf{A}) \text{ is not true} \Longleftrightarrow O(\mathbf{A}) \text{ is true}$$

Since $A(x)$ can be *any* representing formula, we can take it to be $O(x)$ itself. This produces

$$O(\mathbf{O}) \text{ is not true} \Longleftrightarrow O(\mathbf{O}) \text{ is true}$$

which is a contradiction. Hence no such formula $O(x)$ exists.

Thus Russell's paradox provides us with a significant example of a set that is not representable.

4.9 Tarski's Theorem and Gödel's Theorem

Let us examine the argument for the non-representability of the set of ordinary formulas, from the previous section. What would it have taken for the set of ordinary formulas to have been representable? We would have needed to say all of the following.

Russell a $A(x)$ is a representing formula with free variable x.
Russell b $\mathbf{A}$ is a closed term.
Russell c $\mathbf{A}$ designates $A(x)$.
Russell d $A(\mathbf{A})$ is the result of substituting term $\mathbf{A}$ for the free variable in $A(x)$.
Russell e $A(\mathbf{A})$ is not true in the standard model $\mathbb{HF}$.

Let us look at these, one at a time. We have Russell a, it is Rep 26 from Sect. 4.6. Likewise Russell b is Rep 18, Russell c is Rep 21, and Russell d is Rep 27. This leaves Russell e. If we could represent the set of formulas with no free variables that are true in $\mathbb{HF}$, we could get Russell e easily because we have negation available. So, this must be where the problem lies. We know the set of ordinary formulas is not representable. We know it would be if we had representability of Russell a–Russell e. We have representability of Russell a–Russell d, and we would also have Russell e if we had representability of the set of true formulas with no free variables. So it must be that we do not have that. This is a version of

Tarski's Theorem *The set of closed formulas of* LS *that are true in the standard model* $\mathbb{HF}$ *is not representable by a formula of* LS*.*

We discussed the notion of formal axiomatic proof in Sect. 4.6. We noted that, if the non-logical axioms we use constitute a representable set of formulas, the same will be true of the theorems we can prove from them, Rep 28. Tarski's Theorem tells us the set of true formulas is *not* representable. Then no representable set of axioms can ever capture truth. More formally, we have

Gödel's Theorem *The set of closed formulas of* LS *that are true in the standard model* $\mathbb{HF}$ *is not the set of theorems of any first-order axiom system having a representable set of non-logical axioms.*

Oversimplifying things a bit for dramatic purposes, truth for $\mathbb{HF}$ cannot be captured through proof in a formal system. In fact, oversimplification runs rampant here.

4.10 Final Comments

The proof we gave for Gödel's (First) Incompleteness Theorem differs from his in one very significant aspect. Our proof makes essential use of truth; his proof did not. Where we talked about truth in the standard model $\mathbb{HF}$, Gödel talked directly about provability in a formal system of arithmetic. The arithmetic/set difference is not at all essential, but the truth/proof one is. Our argument derives from Tarski, and is not Gödel's proof, though both versions make use of similar ideas.

Suppose we move things in Gödel's direction. Then the notion of ordinary formula needs revision. For us, $A(x)$ was ordinary if $A(\mathbf{A})$ was not true. If we replace truth with provability in a formal system, we have two options: first we could require that $\neg A(\mathbf{A})$ be provable in the formal system, and second, we could require that $A(\mathbf{A})$ be not provable. These are conceptually different, and turn out to be mathematically different as well. As it happens, the first version leads us to a proof of Gödel's Theorem along the constructive lines that Gödel used. The second leads us to Church's Theorem about the undecidability of formal systems of sufficient strength. In both cases the heart of the argument can be based on Russell's paradox. It is even possible to get Rosser's theorem by combining both versions. But all this is way beyond the scope of the present paper whose purpose is purely expository. A full discussion can be found in Fitting (2007).

References

Feferman, S., Dawson Jr., J. W., Kleene, S. C., Moore, G. H., Solovay, R. M., van Heijenoort, J., Goldfarb, W. D., Parsons, C., & Sieg, W. (eds) (1986–2003). *Kurt Gödel collected works*, Five volumes. Oxford: Oxford University Press.

Fitting, M. C. (2007). *Incompleteness in the land of sets*. London: College Publications.

Gödel, K. (1931) Über formal unentscheidbare sätze der *Principia Mathematica* und verwandter systeme I. *Monatshefte für Mathematik und Physik*, 38:173–198. Translated as *On formally undecidable propositions of Principia Mathematica and related systems I*, in Feferman et al. (1986–2003, I, 145–195).

Paulson, L. C. (2014). A machine-assisted proof of Gödel's incompleteness theorems for the theory of hereditarily finite sets. *Review of Symbolic Logic*, *7*(3), 484–498.

Smullyan, R. M. (1956a). Elementary formal systems (abstract). *Bulletin of the American Mathematical Society*, *62*, 600.

Smullyan, R. M. (1956b). On definability by recursion (abstract). *Bulletin of the American Mathematical Society*, *62*, 601.
Smullyan, R. M. (1957). Languages in which self-reference is possible. *Journal of Symbolic Logic*, *22*(1), 55–67.
Smullyan, R. M. (1958). Theories with effectively inseparable nuclei (abstract). *Journal of Symbolic Logic*, *23*(4), 458.
Smullyan, R. M. (1961). *Theory of formal systems* (revised ed.). Princeton, NJ: Princeton University Press.
Smullyan, R. M. (1985). *To Mock a Mockingbird*. Alfred A: Knopf.
Smullyan, R. M. (1987). *Forever undecided*. Alfred A: Knopf.
Smullyan, R. M. (1992). *Gödel's incompleteness theorems*. Oxford: Oxford University Press.
Smullyan, R. M. (1994). *Diagonalization and self-reference*. Oxford: Oxford University Press.
Świerczowski, S. (2003). Finite sets and Gödel's incompleteness theorems. *Dissertationes Mathematicae*.
van Heijenoort, J. (1967). *From Frege to Gödel*. Cambridge, MA: Harvard University Press.
von Neumann, J. (1923). Zur Einfürung der transfiniten Zahlen. *Acta litterarum ac scientiarum Regiae Universitatis Hungaricae Francisco-Josephinae, Sectio scientiarum mathematicarum I*, pp. 199–208. English translation as "On the introduction of transfinite numbers", pp 346–354 of van Heijenoort (1967).
Whitehead, A. N., & Russell, B. (1910/1927). *Principia Mathematica*, 2nd edition, 1927. First edition, 1910. Cambridge: Cambridge University Press.
Zermelo, E. (1908). Untersuchungen über die Grundlagen der Mengenlehre, I. *Mathematische Annalen*, 65:261–281. English translation in van Heijenoort (1967, pp. 201–215).

Chapter 5
Dance of the Starlings

Henk Barendregt, Jörg Endrullis, Jan Willem Klop and Johannes Waldmann

Abstract In this birdwatching paper our binoculars are focused upon a particular bird from Smullyan's enchanted forest of combinatory birds (Smullyan in To Mock a Mockingbird, and other logic puzzles. Alfred A. Knopf, New York, 1985), to wit the Starling. In the feathers of λ-calculus this bird has the plumage $\lambda abc.ac(bc)$. This term is usually named S, reminiscent of its inventor Schönfinkel and also the combinatory ornithologist Smullyan. The combinator S is important for a variety of reasons. First, it is part of the {S, K}-basis for Combinatory Logic (CL). Second, there are several interesting questions and observations around S, mostly referring to termination and word problems. Our paper collects known facts, but poses in addition several new questions. For some of these we provide solutions, but several tough open questions remain.

5.1 Introduction

In the study of important formal systems or structures it is often rewarding to study their subsystems or substructures. Second, it is worthwhile to extend structures, for instance from finitary to infinitary. Both points of view are employed in the present study.

More specifically, in this paper we focus on the subsystem of Combinatory Logic (CL) and λ-calculus, generated by a single term, possibly variables $a, b, c, \ldots$ and using application. We will consider the combinator S with the reduction (or rewrite) rule

H. Barendregt (✉)
Radboud University Nijmegen, Nijmegen, The Netherlands
e-mail: henk@cs.ru.nl

J. Endrullis
VU University Amsterdam, Amsterdam, The Netherlands

J. W. Klop
Centrum Wiskunde en Informatica, Amsterdam, The Netherlands

J. Waldmann
Hochschule Für Technik, Wirtschaft und Kultur (FH) Leipzig, Leipzig, Germany

M. Fitting and B. Rayman (eds.), *Raymond Smullyan on Self Reference*,
Outstanding Contributions to Logic 14, https://doi.org/10.1007/978-3-319-68732-2_5

$$\mathsf{S}\,abc \to ac(bc)$$

The λ-translation is

$$\mathsf{S} \equiv \lambda abc.\,ac(bc)$$

In this paper we presuppose a modest familiarity with the rewriting aspects of Combinatory Logic and λ-calculus; see e.g. Baader and Nipkow (1998), Barendregt (1971, 1984, 1992b), Klop (1980, 1992), Terese (2003), Hindley and Seldin (2008) and Bimbó (2011). Of course there are the classical references to Combinatory Logic and λ-calculus in Schönfinkel (1924), Heijenoort (1967), Church (1941), Curry and Feys (1974), and Curry et al. (1972). For an extensive history of Combinatory Logic and λ-calculus see Cardone and Hindley (2006).

There are several questions presenting themselves.

(1) *Termination.* A typical question that arises is: are there nonterminating S-terms, that is, S-terms without normal form? As is known for a long time, since 1975, there are S-terms with an infinite reduction, or equivalently, without normal form; see Barendregt et al. (1976), Barendregt (1984). Another question solved positively (Waldmann 1998, 2000) much later is about decidability of normalization of S-terms.
(2) *Word problems.* Is the word problem for S-terms solvable? This is still an open question.
(3) *Translations to λ-calculus.* Several questions arise in relation with the translation to λ-calculus (Barendregt 1984). A typical question here is: do S-terms that are in normal form in CL have also a normal form in λ-calculus? Another question is the decidability of strong normalization of S-terms in Combinatory Logic as opposed to that in λ-calculus.
(4) *Infinitary behaviour.* Both CL and λ-calculus have natural infinitary versions, as studied in Terese (2003), Barendregt and Klop (2009), Endrullis et al. (2012a), Kennaway et al. (1997), Kennaway et al. (1995), Endrullis et al. (2010c). The combinator S can be studied in these infinitary calculi: we consider infinite S-terms in infinitary Combinatory Logic (CL^{∞}) and infinitary λ-calculus (λ^{∞}). A typical question here is: does the set of finite S-terms enjoy the property of infinitary strong normalization (SN^{∞})?

In the sequel we will be more precise about these questions.

5.2 Normalization

The combinator S was conceived by Schönfinkel in 1920 (Schönfinkel 1924) together with the combinators K and I. The original names were *Verschmelzung*, *Konstanz* and *Identität*. They are subject to the following rewrite rules:

$$\mathsf{S}xyz \to xz(yz)$$
$$\mathsf{K}xy \to x$$
$$\mathsf{I}x \to x$$

It is well-known that the rewrite system composed of these constants and rules, called Combinatory Logic (CL), enjoys the property of *combinatory completeness* and is able to define all partial recursive functions. The rewrite system has the *confluence* property, also called the *Church-Rosser property*; it is a so-called *orthogonal* rewrite system. Equally well-known is that among the {S, K, I}-terms, also called *combinators*, we have terms that are terminating (or strongly normalizing, SN) and terms with an infinite reduction such as SII(SII) which actually admits a cyclic reduction:

$$\mathsf{SII(SII)} \to \mathsf{I(SII)(I(SII))} \to \mathsf{SII(I(SII))} \to \mathsf{SII(SII)}$$

In a research seminar at Utrecht University in 1974 the question was posed by Jan Bergstra whether the subsystem of CL generated by S only is terminating, in other words the set of S-terms has only terminating S-terms. This question triggered a wager among the seminar participants with a price of 25 Dutch Guilder (11.36 Euros) paid by everyone to whom first solved this termination problem. Henk Barendregt won the contest in February 1974 showing that the term SSS(SSS)(SSS) has an infinite reduction. He wrote his solution on the wallpaper of his house under reconstruction at that time, see Fig. 5.1.

We present a recent upgrade of this proof showing that in fact it has an infinite *head* reduction. The first steps of this reduction are shown in Fig. 5.2.

We see that there are many occurrences of $\mathsf{S}AA$, which we abbreviate by D in Fig. 5.3. By another shorthand $E \equiv \mathsf{S}AD$ we obtain Fig. 5.4.

Theorem 1 (Henk Barendregt 1974; upgraded) *The term*

$$\mathsf{SSS(SSS)(SSS)}$$

has an infinite head reduction. Hence also no normal form, see Klop (1980).

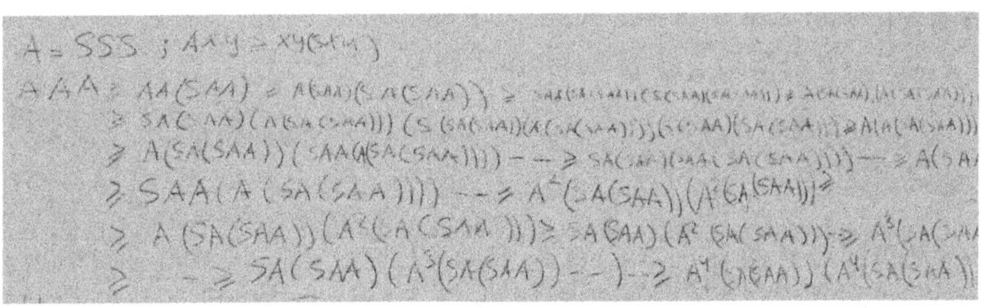

Fig. 5.1 Infinite reduction of AAA written on the wall of Barendregt's house

$$
\begin{array}{l}
AAA \\
\mathsf{SSS}AA \\
\mathsf{S}A(\mathsf{S}A)A \\
AA(\mathsf{S}AA) \\
\mathsf{SSS}A(\mathsf{S}AA) \\
\mathsf{S}A(\mathsf{S}A)(\mathsf{S}AA) \\
A(\mathsf{S}AA)(\mathsf{S}A(\mathsf{S}AA)) \\
\mathsf{SSS}(\mathsf{S}AA)(\mathsf{S}A(\mathsf{S}AA)) \\
\mathsf{S}(\mathsf{S}AA)(\mathsf{S}(\mathsf{S}AA))(\mathsf{S}A(\mathsf{S}AA)) \\
\mathsf{S}AA(\mathsf{S}A(\mathsf{S}AA))(\mathsf{S}(\mathsf{S}AA)(\mathsf{S}A(\mathsf{S}AA))) \\
A(\mathsf{S}A(\mathsf{S}AA))(A(\mathsf{S}A(\mathsf{S}AA)))(\mathsf{S}(\mathsf{S}AA)(\mathsf{S}A(\mathsf{S}AA))) \\
\mathsf{SSS}(\mathsf{S}A(\mathsf{S}AA))(A(\mathsf{S}A(\mathsf{S}AA)))(\mathsf{S}(\mathsf{S}AA)(\mathsf{S}A(\mathsf{S}AA))) \\
\mathsf{S}(\mathsf{S}A(\mathsf{S}AA))(\mathsf{S}(\mathsf{S}A(\mathsf{S}AA)))(A(\mathsf{S}A(\mathsf{S}AA)))(\mathsf{S}(\mathsf{S}AA)(\mathsf{S}A(\mathsf{S}AA))) \\
\mathsf{S}A(\mathsf{S}AA)(A(\mathsf{S}A(\mathsf{S}AA)))(\mathsf{S}(\mathsf{S}A(\mathsf{S}AA))(A(\mathsf{S}A(\mathsf{S}AA))))(\mathsf{S}(\mathsf{S}AA)(\mathsf{S}A(\mathsf{S}AA)))
\end{array}
$$

Fig. 5.2 The head reduction of AAA. Head redexes are underlined red and root redexes are highlighted blue (unmarked steps are definition unfolding steps). After the three root reduction steps as displayed, no further root steps take place: the root has become stable

$$
\begin{array}{l}
\vdots \\
AAD \\
\mathsf{SSS}AD \\
\mathsf{S}A(\mathsf{S}A)D \\
AD(\mathsf{S}AD) \\
\mathsf{SSS}D(\mathsf{S}AD) \\
\mathsf{S}D(\mathsf{S}D)(\mathsf{S}AD) \\
\mathsf{S}AA(\mathsf{S}AD)(\mathsf{S}D(\mathsf{S}AD)) \\
A(\mathsf{S}AD)(A(\mathsf{S}AD))(\mathsf{S}D(\mathsf{S}AD)) \\
\mathsf{SSS}(\mathsf{S}AD)(A(\mathsf{S}AD))(\mathsf{S}D(\mathsf{S}AD)) \\
\mathsf{S}(\mathsf{S}AD)(\mathsf{S}(\mathsf{S}AD))(A(\mathsf{S}AD))(\mathsf{S}D(\mathsf{S}AD)) \\
\mathsf{S}AD(A(\mathsf{S}AD))(\mathsf{S}(\mathsf{S}AD)(A(\mathsf{S}AD)))(\mathsf{S}D(\mathsf{S}AD)) \\
A(A(\mathsf{S}AD))(\mathsf{S}AA(A(\mathsf{S}AD)))(\mathsf{S}(\mathsf{S}AD)(A(\mathsf{S}AD)))(\mathsf{S}D(\mathsf{S}AD)) \\
\mathsf{SSS}(A(\mathsf{S}AD))(\mathsf{S}AA(A(\mathsf{S}AD)))(\mathsf{S}(\mathsf{S}AD)(A(\mathsf{S}AD)))(\mathsf{S}D(\mathsf{S}AD)) \\
\mathsf{S}(A(\mathsf{S}AD))(\mathsf{S}(A(\mathsf{S}AD)))(\mathsf{S}AA(A(\mathsf{S}AD)))(\mathsf{S}(\mathsf{S}AD)(A(\mathsf{S}AD)))(\mathsf{S}D(\mathsf{S}AD))
\end{array}
$$

Fig. 5.3 The head reduction of AAA where $D \equiv \mathsf{S}AA$

Proof In this proof[1, 2], we write $\to$ for head steps. As shorthand define

---

[1]The original proof did not use a leftmost reduction strategy, and required the invocation of a theorem about non-erasing reduction (now known as Church's Theorem (Klop 1980, Theorem 5.9.3) or (Terese 2003, Theorem 4.8.5)) to conclude that the term has no normal form.

[2]See Klop (1980) or see Appendix, Theorem 45, noting that 'no head normal form' implies 'no normal form'.

$$A \equiv \mathsf{SSS} \qquad D \equiv \mathsf{S}AA \qquad E \equiv \mathsf{S}AD \qquad \underline{n} = A^n E\ ,$$

where inductively $A^0E \equiv E$ and $A^{n+1}E \equiv A(A^nE)$ for $n \geq 0$. We define

$$P \rhd Q \iff P \equiv Q\boldsymbol{R}, \text{ for some non-empty vector of } \mathsf{S}\text{-terms } \boldsymbol{R}\ .$$

and $\multimap = (\rightarrow \cup \rhd)$. We claim

(i) $AAA \multimap^+ \underline{1}\,\underline{1}$;
(ii) $\underline{0}\,x \multimap^+ x(Dx) \rhd x$;
(iii) $\underline{n+1}\,x \multimap^+ \underline{n}\,x$;
(iv) $\underline{n}\,x \multimap^* \underline{0}\,x \multimap x$;
(v) $\underline{0}\,\underline{n} \multimap^+ \underline{n+1}\,\underline{n+1}$;
(vi) $\underline{n}\,\underline{n} \multimap^+ \underline{n+1}\,\underline{n+1}$.

From these clauses it follows that $\mathsf{SSS(SSS)(SSS)}$ has an infinite $\multimap$-reduction:

$$AAA \multimap^+ \underline{1}\,\underline{1} \multimap^+ \underline{2}\,\underline{2} \multimap^+ \underline{3}\,\underline{3} \multimap^+ \underline{4}\,\underline{4} \multimap^+ \ldots$$

First note that this $\multimap$-reduction contains an infinite number of $\rightarrow$ steps (as $\rhd$ is well-founded). We have $\rhd \circ \rightarrow\ \subseteq\ \rightarrow \circ \rhd$ and consequently this infinite $\multimap$-path can be transformed into an infinite $\rightarrow$-path:

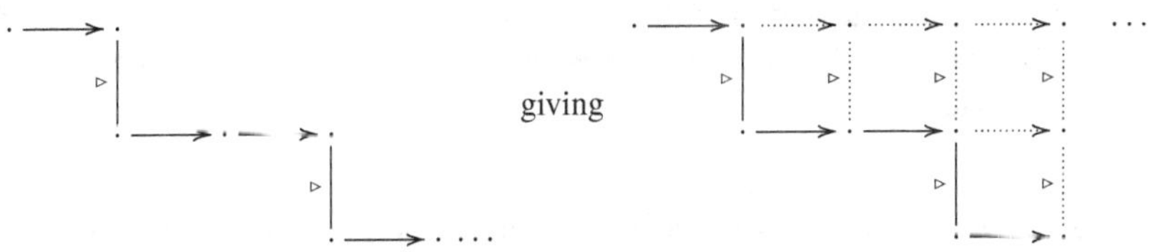

$\vdots$
AAD
$\mathsf{SSS}AD$
$\mathsf{S}A(\mathsf{S}A)D$
ADE
$\mathsf{SSS}DE$
$\mathsf{S}D(\mathsf{S}D)E$
$\mathsf{S}AAE(\mathsf{S}DE)$
$AE(AE)(\mathsf{S}DE)$
$\mathsf{SSS}E(AE)(\mathsf{S}DE)$
$\mathsf{S}E(\mathsf{S}E)(AE)(\mathsf{S}DE)$
$\mathsf{S}AD(AE)(\mathsf{S}E(AE))(\mathsf{S}DE)$
$A(AE)(\mathsf{S}AA(AE))(\mathsf{S}E(AE))(\mathsf{S}DE)$
$\mathsf{SSS}(AE)(\mathsf{S}AA(AE))(\mathsf{S}E(AE))(\mathsf{S}DE)$
$\mathsf{S}(AE)(\mathsf{S}(AE))(\mathsf{S}AA(AE))(\mathsf{S}E(AE))(\mathsf{S}DE)$

Fig. 5.4 The head reduction of AAA where $D \equiv \mathsf{S}AA$ and $E \equiv \mathsf{S}AD$

As S-reduction is non-erasing (Appendix A.2), this infinite head reduction shows that AAA does not have a normal form (Terese 2003).

We now prove the clauses above.

(i) $AAA \equiv \mathsf{SSS}AA \to \mathsf{S}A(\mathsf{S}A)A \to AAD \equiv \mathsf{SSS}AD$
$\to \mathsf{S}A(\mathsf{S}A)D \to AD(\mathsf{S}AD) \equiv \mathsf{SSS}DE \to \mathsf{S}D(\mathsf{S}D)E \to DE(\mathsf{S}DE)$
$\rhd DE \equiv \mathsf{S}AAE \to AE(AE) \equiv \underline{1}\,\underline{1}$

(ii) $\underline{0}x \equiv Ex \equiv \mathsf{S}ADx \to Ax(Dx) \equiv \mathsf{SSS}x(Dx)$
$\to \mathsf{S}x(\mathsf{S}x)(Dx) \to x(Dx)(\mathsf{S}x(Dx)) \rhd x(Dx) \rhd x$

(iii) $\underline{n+1}\,x \equiv A^{n+1}Ex \equiv \mathsf{SSS}(A^nE)x \to \mathsf{S}(A^nE)(\mathsf{S}(A^nE))x$
$\to A^nEx(\mathsf{S}(A^nE)x) \rhd A^nEx \equiv \underline{n}\,x$

(iv) The claim follows by induction using (iii) and (ii).

(v) $\underline{0}\,\underline{n} \multimap^+ \underline{n}(D\underline{n})$ by (ii)
$\multimap^* D\,\underline{n}$ by clause (iv)
$\equiv \mathsf{S}AA\,\underline{n} \to A\,\underline{n}(A\,\underline{n}) \equiv \underline{n+1}\,\underline{n+1}$

(vi) $\underline{n}\,\underline{n} \multimap^* \underline{0}\,\underline{n} \multimap^+ \underline{n+1}\,\underline{n+1}$ by (iv) and (v).

□

Remark 2 Define $\twoheadrightarrow\!\!\rhd\; = \;\to^+ \circ \rhd$, that is, a positive number of head rewrite steps followed by skipping a non-empty vector of arguments. We note that the proof can also be carried out by taking the relation $\twoheadrightarrow\!\!\rhd$ instead of $\multimap^+$: if $\multimap^+$ is replaced by $\twoheadrightarrow\!\!\rhd$ in (i)–(iii) and (v)–(vi), and $\multimap^*$ is replaced by $\twoheadrightarrow\!\!\rhd^*$ in (iv), then all those clauses remain valid. The relation $\twoheadrightarrow\!\!\rhd$ carries the additional information that there is at least one $\to$ step.

We will now present an alternative proof due to den Hartog which is included in Barendregt et al. (1976).

Proof (Proof of Theorem 1 due to den Hartog) Let $\mathcal{H}$ be the set of all passive subterms of all head reducts of AAA. A subterm occurrence N of term M is passive if it does not have an argument. For example, in SSSS, the passive subterms are the second, the third and the fourth S and the entire term.

We show that the following property holds for all terms in $M \in \mathcal{H}$:

(∗) If $M \equiv \mathsf{S}M_1M_2 \dots M_n$, then $M_3, M_4, \dots, M_n$ end in A.

The property (∗) holds for $AAA \equiv \mathsf{SSS}AA$ and we prove its invariance under head reduction. Let M be a term in $\mathcal{H}$ such that all passive subterms of M have the property (∗). We consider a head rewrite step $M \to M'$. Then we have $M \equiv \mathsf{S}M_1M_2 \dots M_n$ and $M' = M_1M_3(M_2M_3)M_4 \dots M_n$. By assumption all passive subterms of $M_1, \dots, M_n$ have property (∗). Thus it suffices to show that the newly formed passive subterms M' and M_2M_3 fulfil (∗):

(i) Let $M_1 \equiv \mathsf{S}N_1 \dots N_m$. Then $M' \equiv \mathsf{S}N_1 \dots N_mM_3(M_2M_3)M_4 \dots M_n$. The arguments $N_3, \dots, N_k$ end in A since M_1 fulfils (∗) and M_3, (M_2M_3), $M_4, \dots, M_n$ end in A since $M_3, \dots, M_n$ do.

(ii) M_2M_3 fulfils $(*)$ since M_2 does and the additional argument M_3 ends in A.

Thus $(*)$ holds for all terms in $\mathcal{H}$.

The reduction of AAA can only terminate in a term S, $\mathsf{S}P$ or $\mathsf{S}PQ$. Clearly, S and $\mathsf{S}P$ have no predecessors. It is a simple exercise to verify that the only possible predecessor of $\mathsf{S}PQ$ is $\mathsf{SS}YP$, and the only possible predecessor of $\mathsf{SS}YP$ is $\mathsf{SS}X\mathsf{S}P$. But this term does not satisfy the condition $(*)$.

Note that the condition $(*)$ determines a regular language. We will use a somewhat similar argument below, in Theorem 4.

In the following Figs. 5.5, 5.6, 5.7 and 5.10 we have drawn some infinite S-terms, and in Fig. 5.15 a finite S-term. Here we used the following drawing conventions. The figures display binary trees; the symbols for the binary application nodes ($\bullet$) are suppressed and appear implicitly at every corner where line segments meet. The terminal nodes are S's; they are colored red. From every line corner (application node) two edges emanate, one blue, one green. Green goes to the first argument M of $M \bullet N$, blue to the second, N. Thus the infinite left-winding spiral in Fig. 5.5, which is green, is the infinite *left spine* of the tree. This is the path starting at the root, in each step progressing to the first argument (which would be drawn on the left in the usual top-down rendering of a term tree). The same in Figs. 5.6 and 5.7.

However, note that in Fig. 5.10 the infinite right-winding spiral is blue; this is not a mistake, but signifies that for this term it is the right spine that is infinite. (The trees in Figs. 5.5, 5.6 and 5.7 have only a finite right spine.)

Figure 5.5 displays the infinite limit of the head reduction of AAA. By Theorem 1, the term AAA admits an infinite reduction

$$AAA \equiv t_0 \to t_1 \to t_2 \to \dots$$

consisting only of head steps. The limit of this reduction is the limit term

$$\lim_{i \to \infty} t_i$$

in the usual metric on terms: $d(s,t) = 0$ if $s \equiv t$ and $d(s,t) = 2^{-n}$ if $s \not\equiv t$ and n is the least depth at which s and t differ. Any infinite reduction starting from a finite S-term has a limit in this sense since every finite prefix eventually becomes stable. This is a consequence of Theorem 29. Taking limits of infinite reductions (and possible reducing further after limit steps) is the core of infinitary rewriting. For further reading on infinitary rewriting we refer the to Dershowitz et al. (1991), Kennaway and de Vries (2003), Endrullis et al. (2012a, 2013a), Endrullis and Polonsky (2013); for a reminder, see Appendix A.5.

There is a *caveat* here. The infinite S-term shown in Fig. 5.5 is not an infinite normal form yet; it contains many redexes. As we have seen, the term AAA has an infinite head reduction. Now a well-known theorem (Barendregt 1984) states that a λ-term (and analogously CL terms) admitting a reduction with infinitely many head reduction steps, does not have a head normal form, see Appendix, Theorem 45. In the

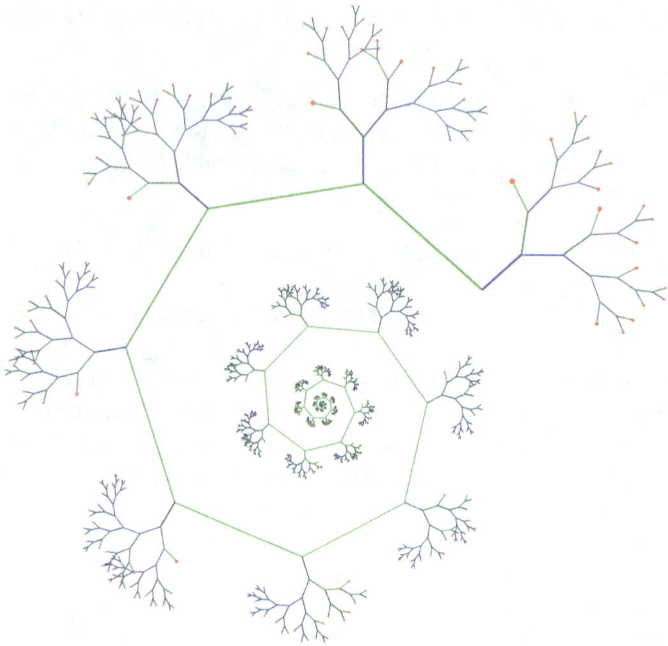

Fig. 5.5 Limit of the infinite head reduction of AAA

case of Combinatory Logic, head normal forms of S-terms are of the form S, SM or SMN. This means that the Böhm tree of AAA is trivial, just consisting of a single $\bot$. There is however another notion of tree, in λ-calculus known as the Berarducci tree (Berarducci 1996; Endrullis et al. 2012a) which records not the progressive stabilisation of head normal forms, but of root normal forms. The limit of the infinite head reduction of AAA can be seen as a partially evaluated Berarducci tree. For more about Böhm trees (BT's) and Berarducci trees (BeT's), their origin and differences, see Appendix A.6.

Figures 5.6 and 5.7 give renderings the totally evaluated Berarducci tree of AAA. This is indeed an infinitary normal form.

We will now consider some more S-terms occurring in the literature of which non-termination has been observed. Alberto Pettorossi has shown in 1975 that the term

$$\mathsf{S}AA(\mathsf{S}AA)$$

has an infinite reduction where $A \equiv \mathsf{SSS}$ (Barendregt 1984; Barendregt et al. 1976, Ex. 7.4.5).

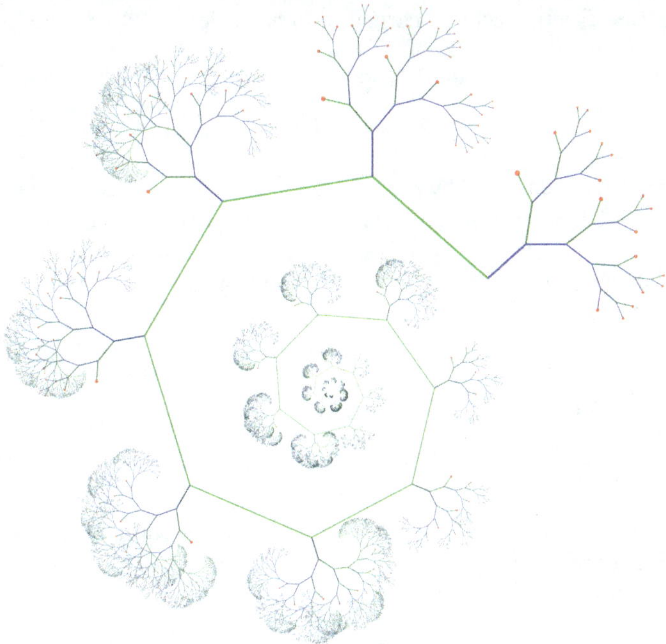

Fig. 5.6 Infinite normal form of AAA (illustration with stretched spine). It is the Berarducci tree of AAA, $\mathrm{BeT}(AAA)$. Note that we encounter on the spine subtrees that may be infinite (1) or finite (0). In fact, the sequence starts with: 0011110011111000111 .... This sequence is computable because normalization (being 0 or 1) is decidable as we will see later

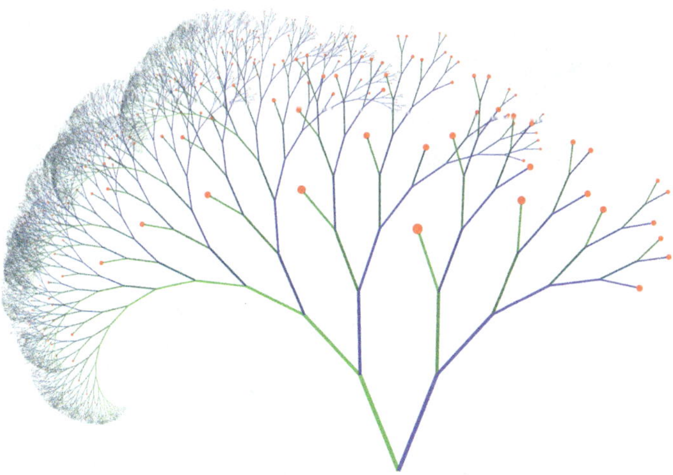

Fig. 5.7 Infinite normal form of AAA, the Berarducci tree of AAA, $\mathrm{BeT}(AAA)$

The shortest S-terms without normal form are of length 7; there are in fact two such terms:

$$X_1 \equiv \mathsf{S(SS)SSSS}$$
$$X_2 \equiv \mathsf{SSS(SS)SS}$$

These terms have been found by Marcel Duboué using exhaustive computer search (Barendregt 1984; Barendregt et al. 1976, Ex. 7.4.5). This search revealed that all shorter S-terms and also all other S-terms of length 7 are normalizing. Monique Baron has shown that X_1 and X_2 are non-normalizing (Barendregt 1984; Barendregt et al. 1976, Ex. 7.4.5).

Further S-terms without normal form are:

$$\mathsf{S}AAA \qquad \mathsf{S}BBB \qquad AA(\mathsf{SS})$$

where $B \equiv \mathsf{S(SS)}$.

Remark 3 For later use (proof of Proposition 6 (ii)) we will now give the normal form of S(SS)(SS)(SS)SS. The normal form is $(\mathsf{S}H)^3(TH)$ which is illustrated as a directed acyclic graph in Fig. 5.8.

The following theorem gives a criterion for an S-term to have an infinite reduction. In particular it states that if M is an S-term of the shape

$$M \equiv C[\ C_1[\mathsf{S}A_1B_1]\ C_2[\mathsf{S}A_2B_2]\ C_3[\mathsf{S}A_3B_3]\ ]$$

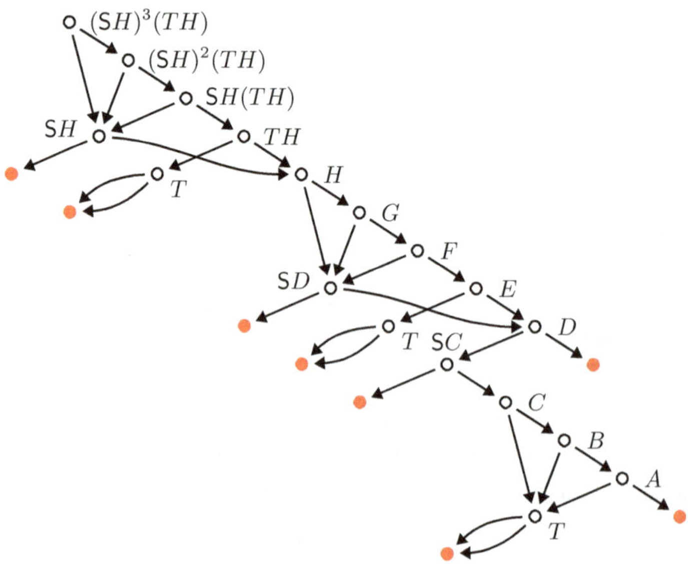

Fig. 5.8 Normal form dag of S(SS)(SS)(SS)SS with $T \equiv \mathsf{SS}$, $H \equiv \mathsf{S}DG$, $G \equiv \mathsf{S}DF$, $F \equiv \mathsf{S}DE$, $E \equiv TD$, $D \equiv \mathsf{SCS}$, $C \equiv TB$, $B \equiv TA$, $A \equiv T\mathsf{S}$

for contexts C, C_i and terms A_i, B_i for $1 \leq i \leq 3$; then M has an infinite reduction.

We introduce some auxiliary notation. We write Σ for the set of all S-terms. For sets of terms P, Q and contexts C we define

$$PQ = \{ MN \mid M \in P, N \in Q \} \qquad C[P] = \{ C[M] \mid M \in P \}$$

For terms M we write $M \to P$ if there exists $N \in P$ such that $M \to N$.

Theorem 4 (Waldmann 2000) *We define a set of* S*-terms*

$$\mathcal{Q} = \{ C[\mathsf{S}MN] \mid \textit{context}\, C,\ \textit{terms}\, M, N \}$$

Every term $M \in \mathcal{QQQ}$ admits an infinite head reduction. It follows that every term $M \in C[\mathcal{QQQ}]$ has an infinite reduction; we will refer to this non-termination criterion as the $\mathcal{QQQ}$-criterion.

Proof It suffices to show that for every term $M \in \mathcal{QQQ}$ there is $N \in \mathcal{QQQ}$ with $M \to N$ via a head step. Let $M \in \mathcal{QQQ}$, that is, $M \equiv M_1 M_2 M_3$ for $M_1, M_2, M_3 \in \mathcal{Q}$. Let Σ_0 be the set of all closed S-terms (Fig. 5.9).

We perform a case distinction on the shape of M_1:

(i) If $M_1 \in \mathsf{S}\mathcal{Q}$, then $M_1 M_2 M_3 \to \mathcal{Q} M_3 (M_2 M_3) \subseteq \mathcal{QQQ}$.
(ii) If $M_1 \in \mathsf{S}\Sigma_0\Sigma_0$, then $M_1 M_2 M_3 \to \Sigma_0 M_2 (\Sigma_0 M_2) M_3 \subseteq \mathcal{QQQ}$.
(iii) If $M_1 \in \mathsf{S}\Sigma_0\Sigma_0\Sigma_0 \ldots$, then $M_1 \to M_1' \in \mathcal{Q}$ and $M_1' M_2 M_3 \in \mathcal{QQQ}$.

In each case, the term M has a head reduct in $\mathcal{QQQ}$. This concludes the proof. □

We will now use this theorem to show non-termination of some of the terms mentioned earlier. The easiest cases are AAA and $\mathsf{S}AAA$ which are immediately in the form $\mathcal{QQQ}$. We emphasize that $\mathcal{QQQ}$ associates to the left. For example, the

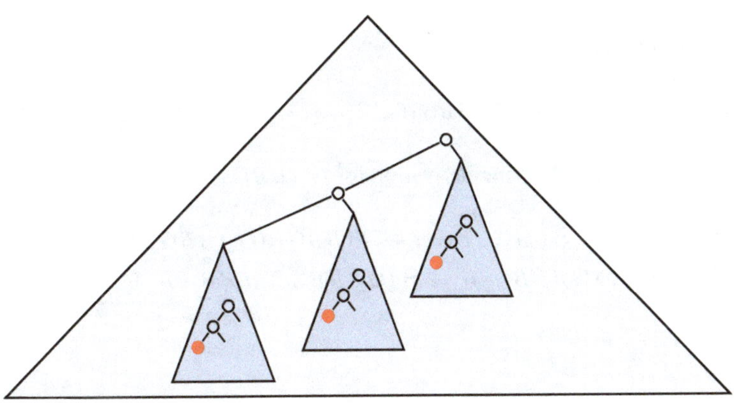

Fig. 5.9 Illustration of the $\mathcal{QQQ}$-criterion (Theorem 4)

term $A(AA)$ is normalizing. Also the Pettorossi term $\mathsf{S}AA(\mathsf{S}AA)$ is of this form:

$$\underbrace{\mathsf{S}A}_{\in\mathcal{Q}}\ \underbrace{A}_{\in\mathcal{Q}}\ \underbrace{(\mathsf{S}AA)}_{\in\mathcal{Q}}$$

We introduce the following auxiliary notation

$$T \equiv \mathsf{SS} \qquad B \equiv \mathsf{S}T \qquad C \equiv B\mathsf{S} \equiv \mathsf{S}T\mathsf{S} \qquad D \equiv \mathsf{S}C\mathsf{S}$$

For $X_1 \equiv \mathsf{S}T\mathsf{SSSS}$ the verification goes as follows. The term admits the following rewrite sequence ending in a term of the shape $C[\mathcal{QQQ}]$:

$$\begin{aligned}
&\mathsf{S}T\mathsf{SSSS} \to T\mathsf{S}T\mathsf{SS} \equiv \mathsf{SSS}T\mathsf{SS} \to \mathsf{S}T(\mathsf{S}T)\mathsf{SS} \to T\mathsf{S}(\mathsf{S}T\mathsf{S})\mathsf{S} \equiv \mathsf{SSS}C\mathsf{S}\\
&\to \mathsf{S}C(\mathsf{S}C)\mathsf{S} \to C\mathsf{S}(\mathsf{S}C\mathsf{S}) \equiv \mathsf{S}T\mathsf{SS}D \to T\mathsf{S}(\mathsf{SS})D \equiv \mathsf{SSS}TD\\
&\to \mathsf{S}T(\mathsf{S}T)D \equiv \mathsf{S}TBD \to TD(BD) \equiv \mathsf{SS}D(BD) \to \mathsf{S}(BD)(\underline{D(BD)})\\
&\equiv \mathsf{S}(BD)(\mathsf{S}C\mathsf{S}(BD)) \to \mathsf{S}(BD)(\underbrace{C}_{\in\mathcal{Q}}\ \underbrace{(BD)}_{\in\mathcal{Q}}\ \underbrace{(\mathsf{S}(BD))}_{\in\mathcal{Q}})
\end{aligned}$$

For $X_2 \equiv \mathsf{SSS}T\mathsf{SS}$ we have $\mathsf{SSS}T\mathsf{SS} \to \mathsf{S}T(\mathsf{S}T)\mathsf{SS}$ and we continue the rewrite sequence as for X_1 above.

For $\mathsf{S}BBB$ we have to have slightly more patience. We start with

$$\begin{aligned}
&\mathsf{S}BBB \to BB(BB) \equiv \mathsf{S}TB(BB)\\
&\to T(BB)(B(BB)) \equiv \mathsf{SS}(BB)(B(BB)) \to \mathsf{S}(B(BB))(\underline{BB(B(BB))})
\end{aligned}$$

Now we abbreviate $G_2 \equiv B(BB)$ and $G_3 \equiv B(B(BB))$. Then we continue with the subterm $BB(B(BB))$:

$$BB(B(BB)) \equiv \mathsf{S}TBG_2 \to TG_2G_3 \equiv \mathsf{SS}G_2G_3 \to \mathsf{S}G_3(\underline{G_2G_3})$$

Now we consider G_2G_3:

$$G_2G_3 \equiv \mathsf{S}T(BB)G_3 \to TG_3(BBG_3) \equiv \mathsf{SS}G_3(BBG_3) \to \mathsf{S}(BBG_3)(\underline{G_3(BBG_3)})$$

Finally, we have the following derivation for $G_3(BBG_3)$:

$$\begin{aligned}
&G_3(BBG_3) \equiv \mathsf{S}TG_2(BBG_3) \to T(BBG_3)(G_2(BBG_3))\\
&\equiv \mathsf{SS}(BBG_3)(G_2(BBG_3)) \to \mathsf{S}\ (G_2(BBG_3))\ (\underbrace{BB}_{\in\mathcal{Q}}\ \underbrace{G_3}_{\in\mathcal{Q}}\ \underbrace{(G_2(BBG_3))}_{\in\mathcal{Q}})
\end{aligned}$$

Let us now prove that AAT has an infinite reduction. Let $H \equiv \mathsf{S}AT$. We have:

$$\begin{aligned} AAT &\equiv \mathsf{SSS}AT \to \mathsf{S}A(\mathsf{S}A)T \to AT(\mathsf{S}AT) \equiv \mathsf{SSS}TH \\ &\to \mathsf{S}T(\mathsf{S}T)H \equiv \mathsf{S}TBH \to TH(BH) \equiv \mathsf{SS}H(BH) \\ &\to \mathsf{S}(BH)(H(BH)) \to \mathsf{S}(BH)(\underline{H(BH)}) \to \mathsf{S}(BH)(\underbrace{A}_{\in Q}\underbrace{(BH)}_{\in Q}\underbrace{(T(BH))}_{\in Q}) \end{aligned}$$

An interesting non-terminating S-term has been found by Zachos (1978):

$$\mathsf{S}TT(\mathsf{S}TT)$$

The reduction graph of this term is a single line; each reduct contains exactly one redex. In fact, this term shows that there exist non-terminating S-terms that do not reduce to a term of the form $C[QQQ]$ of Theorem 4. This can be seen by inspection of the complete reduction graph of this term as given in Waldmann (1998). A characterization of all 'Zachos terms' (reduction graph is an infinite line) can be found in Waldmann (1997). It is worth remarking that there are Zachos terms that fulfill the QQQ-criterion.

Figure 5.10 displays the infinite root normal form of the Zachos term $\mathsf{S}TT(\mathsf{S}TT)$. Remarkably, this term has in contrast to AAA above (and other terms fulfilling the

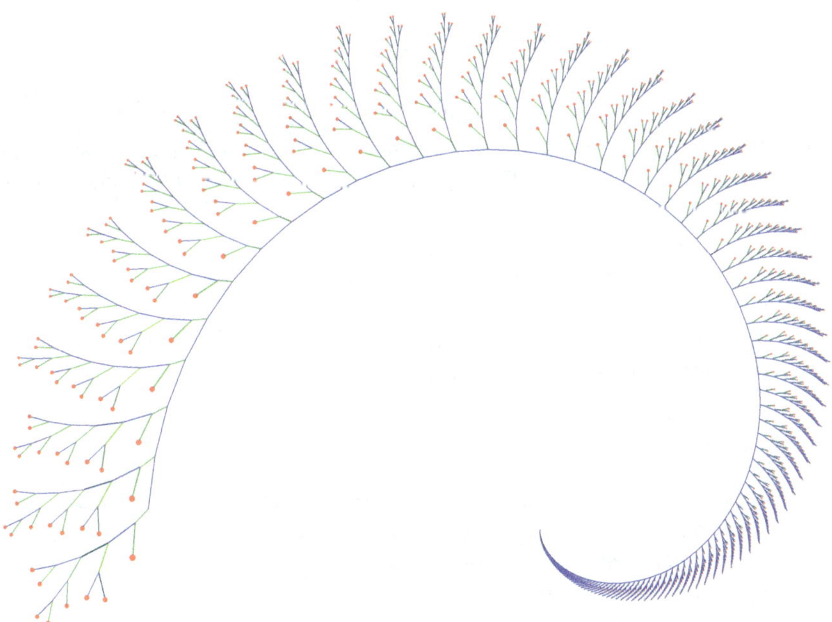

Fig. 5.10 The Berarducci tree (infinite root normal form) of Zachos' term $\mathsf{S}TT(\mathsf{S}TT)$, with $T = \mathsf{SS}$. It has an infinite right spine, blue, and finite left spine, of length 2

$\mathcal{QQQ}$-criterion), next to a fully defined Berarducci tree (infinite root-normal form tree) also a fully defined Böhm tree without $\bot$. We will call a term having a Böhm tree without $\bot$ a *productive* term in analogy with productive definitions of infinite streams as studied in Endrullis et al. (2008, 2010b), Endrullis and Hendriks (2011). So $\mathsf{S}TT(\mathsf{S}TT)$ is productive but AAA is not, as we explained before.

Question 5 How can we characterize the set of productive S-terms? Is it a regular set? Is productivity decidable?

Note that, in general, productivity is an undecidable property (Endrullis et al. 2009c) like most interesting properties of term rewriting systems (Endrullis et al. 2009b, 2011a).

A closer look at the picture of $\mathsf{S}TT(\mathsf{S}TT)$ reveals how the finite subtrees sprouting of the main spiral grow, as a simple computation confirms. This is according to the stream of natural numbers

$$1,\ 2, 2,\ 3, 3, 3,\ 4, 4, 4, 4,\ 5, 5, 5, 5, 5, \ldots$$

Indeed these finite subtrees are of the form $\mathsf{S}Z_i$ where

$$Z_0 \equiv \mathsf{S}TT \qquad\qquad Z_{n+1} \equiv \mathsf{SS}Z_n$$

and now these finite subtrees occur as in the sequence

$$\mathsf{S}Z_1, \mathsf{S}Z_2, \mathsf{S}Z_2, \mathsf{S}Z_3, \mathsf{S}Z_3, \mathsf{S}Z_3, \ldots$$

according to the stream of natural numbers displayed above.

Let us conclude this part about non-termination of S-terms with a question that attracted our attention around 1975. Define for S-terms the *refinement* relation $\lhd$ as follows:

$$C[\mathsf{S}] \lhd C[M], \text{ for a context } C \text{ and a term } M.$$

So $C[M]$ results from $C[\mathsf{S}]$ by replacing one occurrence of S by M. For example, the Zachos' term refines in four steps to the term of Pettorossi:

$$\mathsf{S}TT(\mathsf{S}TT) \lhd^* \mathsf{S}AA(\mathsf{S}AA)$$

Proposition 6 *(i) Let M be an S-term satisfying the $\mathcal{QQQ}$-criterion, and thus be non-terminating. Let $M \lhd N$. Then also N satisfies the $\mathcal{QQQ}$-criterion. So any refinement of a term which is non-terminating by virtue of the $\mathcal{QQQ}$-criterion, is again non-terminating.*

(ii) But in general a refinement of a non-terminating term need not to be non-terminating again.

Proof (i) Let M satisfy the QQQ-criterion

$$M \equiv C[C_1[\mathsf{S}P_1Q_1]C_2[\mathsf{S}P_2Q_2]C_3[\mathsf{S}P_3Q_3]]$$

If the S being refined is in one of the contexts C, C_i $(i = 1, 2, 3)$ the refinement clearly satisfies the same form. Likewise if it is in one of the P_i, Q_i $(i = 1, 2, 3)$. If a head-S of one of the three 'pre-redexes' $\mathsf{S}P_iQ_i$ is refined by $\mathsf{S}R_1 \ldots R_m$ the result clearly contains again a pre-redex.

(ii) We have $\mathsf{SSS}T\mathsf{SS} \lhd^* \mathsf{S}TTT\mathsf{SS}$. Here the left $\mathsf{SSS}T\mathsf{SS}$ was seen above to be non-terminating because one of its reducts satisfies the QQQ-criterion. But the right $\mathsf{S}TTT\mathsf{SS}$ has been shown to be terminating.

The upshot of the proposition above is that terms satisfying the QQQ-criterion such as $\mathsf{SSS}(\mathsf{SSS})(\mathsf{SSS})$ can in fact be seen as *schemes* for non-terminating terms, to the effect that any term $M_1M_2M_3(M_4M_5M_6)(M_7M_8M_9)$ is again non-terminating. Bearing this in mind it seems that there are relatively many non-terminating S-terms compared with the amount of terminating ones. Indeed one can show a zero-one-law to the effect that very large S-terms are with probability almost 1 non-terminating. Actually, such a zero-one law holds for arbitrary context-closed tree languages of S-terms (of which the non-normalizing S-terms are an example).

Question 7 (Colour change puzzle) In this question we will call terminating S-terms *green*, and non-terminating S-terms *red*. We are interested in how during refinements $M \lhd M' \lhd M'' \lhd \ldots$ colours can change from red to green and vice versa. As we have seen, there are red terms that will stay red in every refinement. There is also the red term $\mathsf{SSS}T\mathsf{SS}$ that refines to a green term. Further, it is easy to see that a green term can always be refined to a red term. Now the question: are there terms M with an infinite sequence of refinements

$$M \equiv M_0 \lhd M_1 \lhd M_2 \lhd \ldots$$

in which there are infinitely many colour changes? Or do colours eventually stabilise to red in a refinement sequence?

We expect that the automaton in Fig. 5.11 accepting the terminating (green) S-terms will be helpful to answer this question positively, and even more to characterize this particular class of green terms that can infinitely flip-flop as to colours.

Example 8 (i) The refinement sequence $\mathsf{S} \lhd \mathsf{SS} \lhd \mathsf{SSS} \lhd \mathsf{SSSS} \lhd \ldots$ is entirely green; all terms in this sequence are terminating according to Theorem 18.

(ii) Define the sequence of triangular S-terms by $\Delta_0 = \mathsf{S}$ and $\Delta_{n+1} = \Delta_n\Delta_n$. The sequence $\Delta_0 \lhd \Delta_1 \lhd^* \Delta_2 \lhd^* \Delta_3 \lhd^* \Delta_4 \ldots$ is red from Δ_4 onward.

	0	1	2	3	4	5	6	7	8	9	10	11	12	13	14	15	16	17	18	19	20	21	22	23	24	25	26	27	28	29	30	31	32	33	34	35	36	37
0	1	2	3	4	5	6	6	8	5	10	5	12	5	4	5	10	5	5	5	20	6	6	6	6	6	6	6	6	6	6	6	6	21	6	6	6	21	6
1	7	9	11	13	14	5	6	15	16	15	17	11	18	13	14	15	16	17	18	19	21	21	23	23	24	25	26	27	28	29	30	31	32	33	34	35	36	37
2	19	22	24	25	25	25	25	26	25	26	25	25	25	25	25	26	25	25	25	25	25	25	25	25	25	25	25	25	25	29	30	31	31	31	29	29	31	31
3	27	28	29	29	30	30	30	29	30	29	30	29	30	29	30	29	30	30	30	29	30	30	30	30	30	30	30	30	30	30	30	30	29	30	30	30	29	30
4	31	31	29	29	30	30	30	29	30	29	30	29	30	29	30	29	30	30	30	29	30	30	30	30	30	30	30	30	30	30	30	30	29	30	30	30	29	30
5	29	30	30	30	30	30	30	30	30	30	30	30	30	30	30	30	30	30	30	30	30	30	30	30	30	30	30	30	30	30	30	30	30	30	30	30	30	30
6	30	30	30	30	30	30	30	30	30	30	30	30	30	30	30	30	30	30	30	30	30	30	30	30	30	30	30	30	30	30	30	30	30	30	30	30	30	30
7	9	24	29	30	30	30	30	29	30	29	30	30	30	30	30	29	30	30	30	30	30	30	30	30	30	30	30	30	30	30	30	30	30	30	30	30	30	30
8	32	33	29	29	29	29	29	34	29	34	29	29	29	29	29	34	29	29	29	29	29	29	29	29	29	29	29	29	29	29	30	29	29	29	29	29	29	29
9	15	26	29	29	30	30	30	34	30	35	30	29	30	29	30	35	30	30	30	30	30	30	30	30	30	30	30	30	30	30	30	30	30	30	30	30	30	30
10	36	37	29	29	29	29	29	35	29	35	29	29	29	29	29	35	29	29	29	29	29	29	29	29	29	29	29	29	29	29	30	29	29	29	29	29	29	29
11	19	25	30	30	30	30	30	29	30	29	30	30	30	30	30	29	30	30	30	30	30	30	30	30	30	30	30	30	30	30	30	30	30	30	30	30	30	30
12	37	31	29	29	30	30	30	29	30	29	30	29	30	29	30	29	30	30	30	29	30	30	30	30	30	30	30	30	30	30	30	30	29	30	30	30	29	30
13	27	25	30	30	30	30	30	29	30	29	30	30	30	30	30	29	30	30	30	30	30	30	30	30	30	30	30	30	30	30	30	30	30	30	30	30	30	30
14	31	31	30	30	30	30	30	29	30	29	30	30	30	30	30	29	30	30	30	30	30	30	30	30	30	30	30	30	30	30	30	30	30	30	30	30	30	30
15	15	25	30	30	30	30	30	29	30	29	30	30	30	30	30	29	30	30	30	30	30	30	30	30	30	30	30	30	30	30	30	30	30	30	30	30	30	30
16	32	31	30	30	30	30	30	29	30	29	30	30	30	30	30	29	30	30	30	30	30	30	30	30	30	30	30	30	30	30	30	30	30	30	30	30	30	30
17	36	31	30	30	30	30	30	29	30	29	30	30	30	30	30	29	30	30	30	30	30	30	30	30	30	30	30	30	30	30	30	30	30	30	30	30	30	30
18	37	31	30	30	30	30	30	29	30	29	30	30	30	30	30	29	30	30	30	30	30	30	30	30	30	30	30	30	30	30	30	30	30	30	30	30	30	30
19	24	29	30	30	30	30	30	30	30	30	30	30	30	30	30	30	30	30	30	30	30	30	30	30	30	30	30	30	30	∞	∞	∞	∞	∞	∞	∞	∞	∞
20	29	29	29	29	30	30	30	29	30	29	30	29	30	29	30	29	30	30	30	30	30	30	30	30	30	30	30	30	30	30	30	30	30	30	30	30	30	30
21	29	29	30	30	30	30	30	29	30	29	30	30	30	30	30	29	30	30	30	30	30	30	30	30	30	30	30	30	30	30	30	30	30	30	30	30	30	30
22	29	35	30	30	30	30	30	29	30	29	30	30	30	30	30	29	30	30	30	30	30	30	∞	∞	∞	∞	∞	30	∞	∞	∞	∞	∞	∞	∞	∞	∞	∞
23	29	29	30	30	30	30	30	29	30	29	30	30	30	30	30	29	30	30	30	30	30	30	∞	∞	∞	∞	∞	30	∞	∞	∞	∞	∞	∞	∞	∞	∞	∞
24	30	30	30	30	30	30	30	30	30	30	30	30	30	30	30	30	30	30	30	30	30	30	∞	∞	∞	∞	∞	30	∞	∞	∞	∞	∞	∞	∞	∞	∞	∞
25	30	30	30	30	30	30	30	30	30	30	30	30	30	30	30	30	30	30	30	∞	30	30	∞	∞	∞	∞	∞	∞	∞	∞	∞	∞	∞	∞	∞	∞	∞	∞
26	29	30	30	30	30	30	30	30	30	30	30	30	30	30	30	30	30	30	30	∞	30	30	∞	∞	∞	∞	∞	∞	∞	∞	∞	∞	∞	∞	∞	∞	∞	∞
27	29	30	30	30	30	30	30	30	30	30	30	30	30	30	30	30	30	30	30	30	30	30	30	30	30	30	30	30	30	∞	∞	∞	∞	∞	∞	∞	∞	∞
28	30	29	30	30	30	30	30	30	30	30	30	30	30	30	30	30	30	30	30	∞	30	30	∞	∞	∞	∞	∞	∞	∞	∞	∞	∞	∞	∞	∞	∞	∞	∞
29	30	∞	∞	∞	∞	∞	∞	∞	∞	∞	∞	∞	∞	∞	∞	∞	∞	∞	∞	∞	∞	∞	∞	∞	∞	∞	∞	∞	∞	∞	∞	∞	∞	∞	∞	∞	∞	∞
30	∞	∞	∞	∞	∞	∞	∞	∞	∞	∞	∞	∞	∞	∞	∞	∞	∞	∞	∞	∞	∞	∞	∞	∞	∞	∞	∞	∞	∞	∞	∞	∞	∞	∞	∞	∞	∞	∞
31	30	30	∞	∞	∞	∞	∞	∞	∞	∞	∞	∞	∞	∞	∞	∞	∞	∞	∞	∞	∞	∞	∞	∞	∞	∞	∞	∞	∞	∞	∞	∞	∞	∞	∞	∞	∞	∞
32	26	30	∞	∞	∞	∞	∞	∞	∞	∞	∞	∞	∞	∞	∞	∞	∞	∞	∞	∞	∞	∞	∞	∞	∞	∞	∞	∞	∞	∞	∞	∞	∞	∞	∞	∞	∞	∞
33	34	30	∞	∞	∞	∞	∞	∞	∞	∞	∞	∞	∞	∞	∞	∞	∞	∞	∞	∞	∞	∞	∞	∞	∞	∞	∞	∞	∞	∞	∞	∞	∞	∞	∞	∞	∞	∞
34	35	∞	∞	∞	∞	∞	∞	∞	∞	∞	∞	∞	∞	∞	∞	∞	∞	∞	∞	∞	∞	∞	∞	∞	∞	∞	∞	∞	∞	∞	∞	∞	∞	∞	∞	∞	∞	∞
35	29	∞	∞	∞	∞	∞	∞	∞	∞	∞	∞	∞	∞	∞	∞	∞	∞	∞	∞	∞	∞	∞	∞	∞	∞	∞	∞	∞	∞	∞	∞	∞	∞	∞	∞	∞	∞	∞
36	25	30	∞	∞	∞	∞	∞	∞	∞	∞	∞	∞	∞	∞	∞	∞	∞	∞	∞	∞	∞	∞	∞	∞	∞	∞	∞	∞	∞	∞	∞	∞	∞	∞	∞	∞	∞	∞
37	29	30	∞	∞	∞	∞	∞	∞	∞	∞	∞	∞	∞	∞	∞	∞	∞	∞	∞	∞	∞	∞	∞	∞	∞	∞	∞	∞	∞	∞	∞	∞	∞	∞	∞	∞	∞	∞

Fig. 5.11 Tree automaton accepting the set of strongly normalizing, closed S-terms. More precisely: every S is interpreted as 0, and the table above gives the interpretation for the application symbol '·'. If M has interpretation x and N has interpretation y then the interpretation of MN is given in the intersection of row x and column y

5.3 Automata Methods

We now turn to another result on S-terms by Johannes Waldmann. The proof is decidedly more complicated than the proof of the previous theorems and including a complete proof is beyond the scope of this paper. It may be interesting, however, to give a sketch of the proof.

Theorem 9 (Waldmann 2000) *The set of normalizing, closed* S*-terms is a regular tree language accepted by the automaton shown in Fig. 5.11. Hence strong normalization of closed* S*-terms is decidable in linear time.*

The original proof of this theorem (Waldmann 1998, 2000) was based on an elaborate analysis of reductions involving a by-hand construction of a tree grammar generating the normalizing, closed S-terms. The minimal deterministic tree automaton corresponding to the grammar is shown in Fig. 5.11. Also the termination proof was hand-crafted. We note that the proof in Waldmann (1998, 2000) made use of a computer program to check that one of the complicated case distinctions was indeed exhaustive. In Zachos et al. (2011) a proof is given without the use of computers.

In Endrullis (2009a), Endrullis et al. (2010a) a method has been presented which allows to fully automatically find the tree automaton accepting the set of normalizing terms and prove termination. We will briefly sketch the ideas behind this approach. The Nerode congruence $\sim_L$ for a tree language L is defined by

$$s \sim_L t \iff (\forall C.\, C[s] \in L \iff C[t] \in L) \qquad (5.1)$$

The well-known theorem of Myhill-Nerode states that a tree language L is regular if and only if the number of equivalence classes with respect to $\sim_L$ is finite. We are interested in L being the set of strongly normalizing S-terms. However, using (5.1) to compute the tree automaton (corresponding to $\sim_L$) raises the following problems:

- We don't know (a priori) how to decide membership in L.
- We cannot test an infinite number of contexts C.

We can, however, approximate both problems. We choose numbers $p, q \in \mathbb{N}$ and approximate L by the set of terms L_p that admit reductions of length $\leq p$, and we check only contexts up to size q. With these approximations, the right-hand side of (5.1) is decidable; let us call this decidable approximation $\sim_{p,q}$.

Let $\mathcal{S}_n$ denote the set of S-terms of height $\leq n$. Then we search $n = 1, 2, 3, \ldots$ such that for every $M, N \in T_n$ there exists $O \in \mathcal{S}_n$ such that $MN \sim_{p,q} O$. If this condition is fulfilled, then it induces a tree automaton as follows. We write $\langle M \rangle$ for the equivalence class of M in T_n with respect to $\sim_{p,q}$ (Figs. 5.12 and 5.13).

- The states of the automaton are the equivalence classes $T_n/_{\sim_{p,q}}$.
- The final states are those equivalence classes containing terms in L_n (admitting reductions of length $\leq p$).
- The interpretation of S is $\langle \mathsf{S} \rangle$.
- The interpretation of application is $[\cdot](\langle M \rangle, \langle N \rangle) = \langle MN \rangle$.

This search for n is guaranteed to terminate for any choice of the numbers p, q. However, if p, q are chosen too small, the resulting automaton is not guaranteed to accept the correct language. *So how do we know whether the choice was good?* First, the resulting tree automaton must be a model of the rewrite system since termination is invariant under rewriting. It then can be checked easily that the set of non-accepted terms is closed under rewriting (the model property) and that they always contain a redex (subset of the regular language of terms containing redexes). Second, we can prove *local termination* on all accepted terms by following (Endrullis 2009a). This approach employs semantic labelling (Zantema 1995) of the rewrite system with the states of the tree automaton. Afterwards *global* termination of the labelled system can be proven using automated termination tools. Thus it can be automatically verified that the resulting automaton accepts precisely the terminating S-terms. More details can be found in Endrullis (2009a).

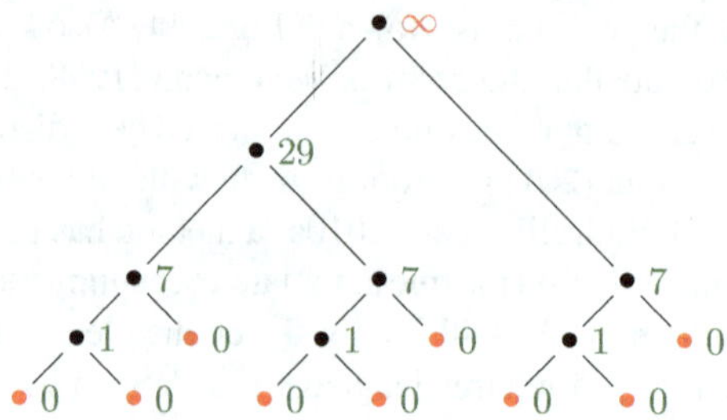

Fig. 5.12 Non-termination of AAA as established by the automaton in Fig. 5.11. The symbol ∞ designates that the term has no interpretation in the automaton (the red area) and hence has an infinite reduction (is non-normalizing)

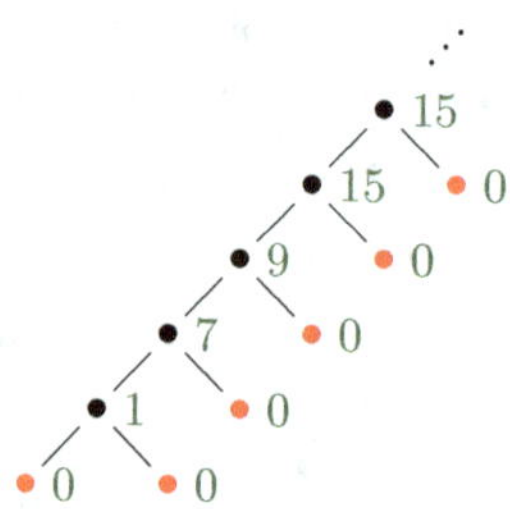

Fig. 5.13 Termination of all flat S-terms established by the automaton in Fig. 5.11. Such terms are normalizing as we will see below in Theorem 18

Question 10 We conjecture that the tree automaton in Fig. 5.14 accepts the closed S-terms with a head normal form. The automaton has been obtained using the construction described above (but now L is the language of head normalizing terms instead of normalizing terms).

As this construction is an approximation, it is crucial to verify that the obtained automaton indeed accepts the expected language.

It is easy to verify that the automaton is a model and hence its language and complement are closed under rewriting.

Moreover we expect that it is not difficult to show that every term with interpretation ♂ contains a head redex (regular language inclusion), and as a consequence admits infinite head reductions.

The challenge is to verify that all terms with interpretations other than ♂ have a head normal form. Can this *local* head normalization problem be verified automatically? Is it possible to reduce this problem to a standard termination problem using semantic labelling similar to the construction in Endrullis (2009a)?

	0	1	2	3	4	5	6	7	8	9	10	11	12	13	14	15	16	17	18	19	20	21	22	23	24	25	26	27	28	29	30	31	32	33	♂
0	1	2	3	4	5	5	7	8	9	10	10	12	8	12	15	8	8	9	19	9	21	9	10	10	10	10	27	10	9	9	9	27	9	10	10
1	6	11	13	13	13	13	13	13	13	13	13	13	13	13	13	13	13	13	13	13	13	13	13	13	13	13	13	13	13	13	13	13	13	13	13
2	14	16	8	9	10	10	17	9	10	10	10	17	9	17	10	10	10	10	10	10	10	10	10	10	10	10	10	10	10	10	10	10	10	10	10
3	18	20	20	20	20	20	20	20	20	20	20	20	20	20	20	20	20	20	20	20	20	20	20	20	20	20	20	20	20	20	20	20	20	20	20
4	8	10	10	10	10	10	10	10	10	10	10	10	10	10	10	10	10	10	10	10	10	10	10	10	10	10	10	10	10	10	10	10	10	10	10
5	10	10	10	10	10	10	10	10	10	10	10	10	10	10	10	10	10	10	10	10	10	10	10	10	10	10	10	10	10	10	10	10	10	10	10
6	11	8	20	10	10	10	22	23	23	23	23	22	23	22	24	23	23	23	25	23	23	23	23	23	23	23	25	23	23	23	23	25	23	23	23
7	26	28	29	22	24	24	30	22	24	24	24	30	22	30	24	24	24	24	24	24	24	24	24	24	24	24	24	24	24	24	24	24	24	24	24
8	24	23	23	23	23	23	23	23	23	23	23	23	23	23	23	23	23	23	23	23	23	23	23	23	23	23	23	23	23	23	23	23	23	23	23
9	25	23	23	23	23	23	23	23	23	23	23	23	23	23	23	23	23	23	23	23	23	23	23	23	23	23	23	23	23	23	23	23	23	23	23
10	23	23	23	23	23	23	23	23	23	23	23	23	23	23	23	23	23	23	23	23	23	23	23	23	23	23	23	23	23	23	23	23	23	23	23
11	13	17	20	10	10	10	30	23	23	23	23	29	23	29	24	23	23	23	25	23	23	23	23	23	23	23	25	23	23	23	23	25	23	23	23
12	31	29	29	22	24	24	29	22	24	24	24	29	22	29	24	24	24	24	24	24	24	24	24	24	24	24	24	24	24	24	24	24	24	24	24
13	13	10	20	10	10	10	24	23	23	23	23	24	23	24	24	23	23	23	25	23	23	23	23	23	23	23	25	23	23	23	23	25	23	23	23
14	8	20	10	10	10	10	23	23	23	23	23	23	23	23	23	23	23	23	23	23	23	23	23	23	23	23	23	23	23	23	23	23	23	23	23
15	24	24	24	24	24	24	24	24	24	24	24	24	24	24	24	24	24	24	24	24	24	24	24	24	24	24	24	24	24	24	24	24	24	24	24
16	22	29	24	24	24	24	24	24	24	24	24	24	24	24	24	24	24	24	24	24	24	24	24	24	24	24	24	24	24	24	24	24	24	24	24
17	22	23	23	23	23	23	23	23	23	23	23	23	23	23	23	23	23	23	23	23	23	23	23	23	23	23	23	23	23	23	23	23	23	23	23
18	20	24	23	23	23	23	23	23	23	23	23	23	23	23	23	23	23	23	23	23	23	23	23	23	23	23	23	23	23	23	23	23	23	23	23
19	32	25	25	25	25	25	25	25	25	25	25	25	25	25	25	25	25	25	25	25	25	25	25	25	25	25	25	25	25	25	25	25	25	25	25
20	23	24	23	23	23	23	23	23	23	23	23	23	23	23	23	23	23	23	23	23	23	23	23	23	23	23	23	23	23	23	23	23	23	23	23
21	33	23	23	23	23	23	23	23	23	23	23	23	23	23	23	23	23	23	23	23	23	23	23	23	23	23	23	23	23	23	23	23	23	23	23
22	25	23	23	23	23	23	♂	♂	♂	♂	♂	♂	♂	♂	♂	♂	♂	♂	♂	♂	♂	♂	♂	♂	♂	♂	♂	♂	♂	♂	♂	♂	♂	♂	♂
23	♂	♂	♂	♂	♂	♂	♂	♂	♂	♂	♂	♂	♂	♂	♂	♂	♂	♂	♂	♂	♂	♂	♂	♂	♂	♂	♂	♂	♂	♂	♂	♂	♂	♂	♂
24	23	23	23	23	23	23	♂	♂	♂	♂	♂	♂	♂	♂	♂	♂	♂	♂	♂	♂	♂	♂	♂	♂	♂	♂	♂	♂	♂	♂	♂	♂	♂	♂	♂
25	23	♂	♂	♂	♂	♂	♂	♂	♂	♂	♂	♂	♂	♂	♂	♂	♂	♂	♂	♂	♂	♂	♂	♂	♂	♂	♂	♂	♂	♂	♂	♂	♂	♂	♂
26	17	23	23	23	23	23	♂	♂	♂	♂	♂	♂	♂	♂	♂	♂	♂	♂	♂	♂	♂	♂	♂	♂	♂	♂	♂	♂	♂	♂	♂	♂	♂	♂	♂
27	25	25	25	25	25	25	25	25	25	25	25	25	25	25	25	25	25	25	25	25	25	25	25	25	25	25	25	25	25	25	25	25	25	25	25
28	30	23	23	23	23	23	♂	♂	♂	♂	♂	♂	♂	♂	♂	♂	♂	♂	♂	♂	♂	♂	♂	♂	♂	♂	♂	♂	♂	♂	♂	♂	♂	♂	♂
29	24	23	23	23	23	23	♂	♂	♂	♂	♂	♂	♂	♂	♂	♂	♂	♂	♂	♂	♂	♂	♂	♂	♂	♂	♂	♂	♂	♂	♂	♂	♂	♂	♂
30	29	23	23	23	23	23	♂	♂	♂	♂	♂	♂	♂	♂	♂	♂	♂	♂	♂	♂	♂	♂	♂	♂	♂	♂	♂	♂	♂	♂	♂	♂	♂	♂	♂
31	10	23	23	23	23	23	♂	♂	♂	♂	♂	♂	♂	♂	♂	♂	♂	♂	♂	♂	♂	♂	♂	♂	♂	♂	♂	♂	♂	♂	♂	♂	♂	♂	♂
32	24	23	♂	♂	♂	♂	♂	♂	♂	♂	♂	♂	♂	♂	♂	♂	♂	♂	♂	♂	♂	♂	♂	♂	♂	♂	♂	♂	♂	♂	♂	♂	♂	♂	♂
33	♂	23	♂	♂	♂	♂	♂	♂	♂	♂	♂	♂	♂	♂	♂	♂	♂	♂	♂	♂	♂	♂	♂	♂	♂	♂	♂	♂	♂	♂	♂	♂	♂	♂	♂
♂	♂	♂	♂	♂	♂	♂	♂	♂	♂	♂	♂	♂	♂	♂	♂	♂	♂	♂	♂	♂	♂	♂	♂	♂	♂	♂	♂	♂	♂	♂	♂	♂	♂	♂	♂

Fig. 5.14 Does this tree automaton accept the set of head normalizing, closed S-terms? Every S is interpreted as 0, and the table above gives the interpretation for the application symbol. All states, except state ♂ are accepting (♂ represents terms without a head normal form). Note that this symbol is also a state of the automaton, it also occurs in the header of the table. This must be so, since having an infinite head reduction is not closed under contexts (in contrast with the property 'having an infinite reduction')

We will now mention the smallest S-terms without a head normal form:

S(SSS)S(SSS)	(SSS)(S(SSS))S	S(SS)S(SSS)S
S(S(SSS))(SS)S	(SSS)(SS)(SS)S	S(SSS)S(SS)S
S(SS)SS(SS)S	(SSS)(S(SS))SS	S(S(SS))(SS)SS
S(SS)S(SS)SS	S(S(SSS))SSS	S(S(SS)S)SSS
(SSS)(SS)SSS	S(S(SS))SSSS	S(SS)SSSSS

All these terms have size 8. We illustrate the working of the automaton in Fig. 5.14 at the example of the first term. To save some space we will this time render the state propagation directly in the term in a self-explanatory manner:

$$\Big(\; ((\,\mathsf{S}_0\; ((\mathsf{S}_0\mathsf{S}_0)_1\mathsf{S}_0)_6\;)_7\; \mathsf{S}_0)_{26}\; ((\mathsf{S}_0\mathsf{S}_0)_1\mathsf{S}_0)_6 \; \Big)_{\infty}$$

To see why ∞ reoccurs in the header of the table in Fig. 5.14, consider the example just treated but now prefixed with S:

$$(\,\mathsf{S}_0\;\; (\ldots)_{\infty}\;)_{10}$$

5.4 Periodic Terms and Reductions

In this section we will investigate cyclic reductions (and variants) of S-terms.

Definition 11 We define three forms of periodic reductions:

(i) The simplest form of a periodic reduction sequence is that of a *cyclic* reduction $M \to \ldots \to M$ for a term M.
(ii) A more general notation is that of a *spiralling* reduction $M \to \ldots \to C[M]$ for a context C. If the context is trivial, $C \equiv \Box$, this is a cyclic reduction.
(iii) An even more general notion is that of an *open spiralling* reduction where M is an open term (possibly containing variables) and $M \to \ldots \to C[M\sigma]$ for some context C and substitution σ.

We say that a term (or a rewrite system) is cyclic, spiralling or open spiralling if it admits a reduction with the respective property.

So we distinguish cyclic terms, spiralling terms and open spiralling terms. We have the implications

$$M \text{ is not open spiralling} \quad\Longrightarrow\quad M \text{ is not spiralling} \quad\Longrightarrow\quad M \text{ is not cyclic}$$

or equivalently: cyclic $\Longrightarrow$ spiralling $\Longrightarrow$ open spiralling.

Remark 12 We have

$$\text{SN} \Longrightarrow \text{not open spiralling} \Longrightarrow \text{not spiralling} \Longrightarrow \text{not cyclic}$$

and also, for orthogonal rewrite systems, WN $\Longrightarrow$ not cyclic (see further Terese 2003). Note that Combinatory Logic is an orthogonal term rewriting system.

Clearly, also SN $\Longrightarrow$ WN. One may ask whether already WN $\Longrightarrow$ not Spiralling. But this is not the case. The following is an orthogonal TRS R which is WN but has spirals: $R = \{A \to B(A),\, B(x) \to C\}$.

For more about looping and spiralling infinite reductions we refer to Geser and Zantema (1999); they give examples of term and string rewrite systems that are non-terminating and not spiralling.

Around 1975 Bergstra established by a simple, elegant argument that S-terms do not admit cyclic reductions. We will also have some alternative proofs of the acyclicity of S-terms.

Theorem 13 (Bergstra 1975) *There is no cyclic reduction of* S*-terms.*

Proof Define the *length* of a CL-term M, notation $|M|$, as the number of occurrences of S, K, I in it, so $|\mathsf{S}| = |\mathsf{K}| = |\mathsf{I}| = 1$ and $|MN| = |M| + |N|$. Define the *weight* of a CL-term M, notation $\|M\|$, as follows: $\|\mathsf{S}\| = \|\mathsf{K}\| = \|\mathsf{I}\| = 1$ and $\|\mathsf{MN}\| = 2\|\mathsf{M}\| + \|\mathsf{N}\|$. It is easy to prove that for S-terms:

(i) If $M \to N$ then $|M| \leq |N|$.
(ii) If $M \to N$ and $|M| = |N|$, then the contracted redex must be an S-redex of the form $\mathsf{S}AB\mathsf{S}$ for some A, B.
(iii) For every S-context C (i.e., an S-term with one open place [ ]) we have that $\|M\| > \|N\|$ implies $\|C[M]\| > \|C[N]\|$.

Now suppose a reduction cycle consisting of S-terms exists; let it be $M_0 \to M_1 \to \ldots \to M_n = M_0$ $(n \geq 1)$. According to (i) we have $|M_0| = |M_1| = \ldots = |M_n|$. According to (ii) every redex contracted in this cyclic reduction has the form $\mathsf{S}AB\mathsf{S}$. However, $\|\mathsf{S}AB\mathsf{S}\| = 4\|A\| + 2\|B\| + 9$ while $\|A\mathsf{S}(B\mathsf{S})\| = 4\|A\| + 2\|B\| + 3$. But then we have by item (iii) $\|M_0\| > \|M_1\| > \ldots > \|M_n\| = \|M_0\|$, contradiction. □

This result (Theorem 13) also follows directly from Waldmann's top-termination theorem (Theorem 29, below, together with Theorem 48 in the Appendix).

We give an alternative proof for Theorem 13.

Proof (Alternative proof of Theorem 13) We use a minimal counterexample argument.

Let $\mathcal{C}$ be a cycle of S-terms which has a minimal total number of S's occurring in it. (So that is the sum of the lengths of the S-terms making up the cycle $\mathcal{C}$.)

(i) *Claim 1*: $\mathcal{C}$ contains a root step $SABC \to AC(BC)$.

Suppose not, so the root application symbol is stable during all of the cycle $\mathcal{C}$. Let $\mathcal{C} \equiv M_1 \equiv A_1B_1 \to A_2B_2 \to \ldots \to A_nB_n \equiv A_1B_1 \equiv M_1$. Because the root symbol is stable, $\mathcal{C}$ contains a cycle $\mathcal{C}'$ either in the A_i-part, or in the B_i-part. That cycle $\mathcal{C}'$ is smaller with respect to the number of S's than $\mathcal{C}$, in contradiction with the minimality assumption. This proves Claim 1.

Remark: Actually, this root step $\mathsf{S}ABC$ is of the form $\mathsf{S}AB\mathsf{S}$, since every redex in $\mathcal{C}$ must be of that form, due to length considerations; the length cannot increase during a cycle. This special form $\mathsf{S}AB\mathsf{S}$ for each redex in $\mathcal{C}$ is used in Bergstra's proof, but is not used in the present proof.

(ii) *Claim 2*. Let rs(M) be the length of the *right spine* of the S-term M. So

$$\mathrm{rs}(\mathsf{S}) = 0$$
$$\mathrm{rs}(MN) = 1 + \mathrm{rs}(N)$$

Then rs cannot decrease in an S-step, and is incremented by 1 in a root step. (And in fact by every redex contraction which has the root of the redex lying on the right spine, if and only if. We will not need that information however.)
For example, $\mathsf{S}AB\mathsf{S}$ has rs 1, and its contractum $A\mathsf{S}(B\mathsf{S})$ has has rs 2.
The proof of Claim 2 is an easy exercise; see further the proof of Theorem 14 where we use this property to show the absence of ground spirals.

From Claims 1 and 2 it follows immediately that there cannot be an S-cycle.

The following theorem improves the earlier theorem about the absence of cyclic reductions to absence of spiralling reductions. Note that every cyclic reduction gives rise to a ground spiral, that is without variables (by replacing all variables by S).

Theorem 14 (Waldmann 2000) *There exists no spiralling* S*-term.*

Proof We define a function $\rho(\cdot)$ on S-terms measuring the length of their right-spine as follows:

$$\rho(x) = 1 \qquad \rho(\mathsf{S}) = 1 \qquad \rho(MN) = 1 + \rho(N)$$

We now label application symbols with length of their right-spine:

$$\overline{x} = x \qquad \overline{\mathsf{S}} = \mathsf{S} \qquad \overline{M \cdot N} = \overline{M} \cdot^{\rho(N)} \overline{N}$$

We now consider the following S-rule:

$$\mathsf{S} \cdot^{a} x \cdot^{b} y \cdot^{c} z \to (x \cdot^{c} z) \cdot^{1+c} (y \cdot^{c} z) \tag{5.2}$$

for $a, b, c \in \mathbb{N}$, together with the label-lifting rule:

$$x \cdot^{n} y \to x \cdot^{n+1} y \tag{5.3}$$

for $n \in \mathbb{N}$. For all S-terms M, N we have

$$M \to_{\mathsf{S}} N \quad \Longrightarrow \quad \overline{M} \to_{(2)} \to^{*}_{(3)} \overline{N} \tag{5.4}$$

The $\to^{*}_{(3)}$ might be needed to increase labels in the context of the step.
Assume that there is a spiralling reduction $M \to^{+} C[M]$. Then there exists also a spiralling labelled reduction $\overline{M} \to^{+} \overline{C[M]}$ by (5.4). The term $\overline{M}$ is a subterm of $\overline{C[M]}$ and hence we could continue reducing forever using only a bounded set of labels.

As a consequence, it suffices to show the rules (5.2) and (5.3) are strongly normalizing when bounding the labels. Let $m \in \mathbb{N}$ and consider the rewrite system R_m consisting of all instances of rules (5.2) and (5.3) that contain only labels $\leq m$. The system R_m is terminating by the recursive path order with the order $\cdot^a < \cdot^b$ if $a > b$. □

Remark 15 A consequence of the no-spiral theorem is that there is no reducing fixed point combinator among the S-terms. Actually, Statman proved the much more general fact that no proper, regular combinator P can be applicatively composed to yield a fixed point combinator (not just reducing). Here P is *proper* if its reduction rule has the form $Px_1 \dots x_n \to w$ with the right-hand side an applicative combination of $x_1, \dots, x_n$; and P is *regular* if w is of the form $x_1 w'$. These notions are defined in Raymond (1994, p. 319). So S is proper and regular.

Remark 16 The acyclicity of S-terms has as immediate consequence that terms in an infinite reduction have unboundedly increasing length. Indeed, feeding a term such as AAA in a calculator for λ-terms or CL-terms, we witness the length-explosion as in Fig. 5.2. In fact, in any infinite reduction also the size of the contracted redexes grows unboundedly. This follows from the proof of the no-spiral theorem.

Conjecture 17 There are no open spiralling S-terms at all.

A corollary of the acyclicity of S-terms is that a certain class of S-terms is strongly normalizing. This actually extends to the following class of {S, K}-terms defined as follows: call a CL-term *flat*, if it has in the usual notation of application to the left no visible brackets. Formally,

- I, K and S are flat, and
- if M is flat, then MI, MK and MS are flat.

Theorem 18 (Klop 1976) *Flat {S, K}-terms are strongly normalizing.*

Proof We prove SN for a class SP of CL-terms that includes the flat {S, K}-terms defined as follows:

(i) $\mathsf{S}, \mathsf{K} \in \mathrm{SP}$;
(ii) $M \in \mathrm{SP} \implies M\mathsf{S}, M\mathsf{K} \in \mathrm{SP}$;
(iii) $M \in \mathrm{SP} \implies \mathsf{KK}M, \mathsf{KS}M, \mathsf{SK}M, \mathsf{SS}M \in \mathrm{SP}$.

It is easy to check that SP is closed under reduction. One also easily checks that if $M, N \in \mathrm{SP}$ and $M \to N$, we have $|M| = |N|$ in case an S-redex is contracted, and $|M| > |N|$ in case a K-redex is contracted. Now suppose that not all terms in SP are SN. From the last observation it is clear that an infinite reduction of SP-terms $M_0 \to M_1 \to \dots$ must contain a cycle $M_i \to M_{i+1} \to \dots \to M_{i+j+1} = M_i$ for some $i, j \geq 0$. In this cycle only S-redexes can be contracted. But then we can find a cycle of S-terms by replacing in $M_i \to M_{i+1} \to \dots \to M_{i+j+1}$ all occurrences of K by occurrences of S. This contradicts the preceding theorem. So we have proved that all SP terms are SN, in particular flat {S, K}-terms. □

Remark 19 This termination property does not hold for flat {S, K, I}-terms:

$$\begin{aligned} \mathsf{SISSSII} &\rightarrow \mathsf{IS(SS)SII} \rightarrow \mathsf{S(SS)SII} \rightarrow \mathsf{SSI(SI)I} \\ &\rightarrow \mathsf{S(SI)(I(SI))I} \rightarrow \mathsf{S(SI)(SI)I} \rightarrow \mathsf{SII(SII)} \end{aligned}$$

We conclude this subsection mentioning a fact concerning reduction cycles in CL.

To this end, we introduce the notion of a *pure* cycle: this is a cycle $M \rightarrow \ldots \rightarrow M$ such that the reduction graph $\mathcal{G}(M)$ of M consists precisely of this cycle – one could say the cycle is deterministic, no splitting of reduction occurs (every term contains just one redex).

Theorem 20 (Klop 1981)

(i) Lambda calculus has pure cycles of any length.
(ii) By contrast, CL based on I, K, S *has no pure cycles. If M is cyclic, then $\mathcal{G}(M)$ is infinite.*

In the same vein we can define a *pure line*. This is a reduction graph $\mathcal{G}(M)$ which consists of a single line. So every term in $\mathcal{G}(M)$ has only one redex. Above, we have considered an S-term establishing a pure line as its reduction graph.

5.5 The Order of S-Terms

Another noteworthy aspect of S-terms is that they can consume only a limited amount of food.

Definition 21 A CL-term M has *order* $n \in \mathbb{N}$ if

- $Mx_0x_1 \ldots x_{n-1}x_n \rightarrow^* N$ implies that $N \equiv N'x_n$ where x_n does not occur in N', and
- n is minimal with this property.

M has *finite order* if it has order n for some n, and *infinite order* otherwise.

For a more elaborate exposition of the notion of order, including its relation to the undefinedness notions mute and unsolvable, see the Appendix; there some alternative definitions of order are mentioned.

We will now establish that the order of S-terms is 0, 1, 2, or 3. An example of an S-term with order 0 is AAA; for, this term has an infinite head reduction, and hence has no head normal form, which is necessary to consume one or more variables offered as arguments. Examples of order 1, 2 and 3 are easy: e.g. SSS, SS, S respectively.

Remarkably, order 4 is impossible for S-terms.

Theorem 22 *(Visser and van Oostrom 1975, 2002). The order of an* S*-term is* ≤ 3.

The proof by Visser (1975) and van Oostrom (2002),[3] personal communication, is like an elegant chess combination.

We will now present the proof. Let us consider an S-term M followed by 4 variables, $Mxyzu$. Clearly, the u can only be consumed in $Mxyzu$, if $Mxyz$ reduces to a term of the form $\mathsf{S}AB$. This proves to be impossible. Some experiments with M hint to the presence in reducts of $Mxyz$, of a particular kind of terms, namely those with a variable (and not S) *at the end of the right spine*; and applicative combinations of such '*variable-tail-terms*'.

Example 23 We give some examples of terms and normalize them to applicative combinations of variable-tail-terms (these terms are called $\mathcal{T}$ in Definition 24), the latter indicated by underlining.

(i) $\mathsf{SS}xyz \to \mathsf{S}y(xy)z \to \underline{yz(xyz)}$.
(ii) $\mathsf{SSS}xyz \to \mathsf{S}x(\mathsf{S}x)yz \to \underline{xy(\mathsf{S}xy)}\underline{z}$.
(iii) $\mathsf{S}(\mathsf{SS})xyz \to \mathsf{SS}y(xy)z \to \mathsf{S}(xy)(y(xy))z \to \underline{xyz}(\underline{y(xy)z})$.

It turns out that such sets of terms have strong closure properties with respect to reduction. Thus we are led to the following definition. The first item is a general notation.

Definition 24 (i) Let $X, Y \subseteq \mathit{Ter}(\mathrm{CL})$. Then $XY = \{MN \mid M \in X, N \in Y\}$.
(ii) Let $\mathcal{T}$ be the set of (possibly open) S-terms with the right spine ending in a variable $x, y, z \ldots$; inductively defined:

(a) $\mathcal{V} \subseteq \mathcal{T}$;
(b) if $s \in \mathit{Ter}(\mathrm{CL})$ and $t \in \mathcal{T}$, then $st \in \mathcal{T}$.

Note that $\mathsf{S} \notin \mathcal{T}$.

Proposition 25 *The sets* $\mathcal{T}, \mathcal{TT}, \mathcal{TTT}, \ldots, \mathcal{T}^{\sim n}, \ldots$ *are closed under reduction for all* n.

Proof Actually, we need only the first three sets, $\mathcal{T}$, $\mathcal{TT}$ and $\mathcal{TTT}$.

(i) *Closure of* $\mathcal{T}$: for a variable this is trivially true. If $st \to st'$, the induction hypothesis suffices. If $st \to s't$, it is trivial, since s' is again an S-term. The interesting case is when $st \equiv \mathsf{S}ABt$; so the reduction step is a root step. Then $\mathsf{S}ABt \to At(Bt)$, of which the right spine covers the right spine of t, ending in a variable. So $At(Bt) \in \mathcal{T}$.
(ii) *Closure of* $\mathcal{TT}$: consider $MM' \in \mathcal{TT}$. The case of an internal (non-root) step in one of the M, M' is easy, by the closure of $\mathcal{T}$ and the definition of $\mathcal{TT}$. The case of a root step: $MM' \equiv \mathsf{S}ABM' \to AM'(BM')$: Note that if $\mathsf{S}AB \in \mathcal{TT}$, then $\mathsf{S}A \in \mathcal{T}$ and $B \in \mathcal{T}$, and also $A \in \mathcal{T}$. Hence $AM' \in \mathcal{T}$ and $BM' \in \mathcal{T}$, and $AM'(BM') \in \mathcal{TT}$.

[3]The observation of this fact was in 1975 by Albert Visser, who included it in his Ph.D.-dissertation (Visser 1981), in the list of Dutch language 'stellingen' (statements to be defended). He had at that time devised a proof that unfortunately got lost several years later while changing houses. Vincent van Oostrom, hearing about the lost proof from Albert, solved the problem after work in the train home, from Utrecht to Amsterdam (about 20 km distance, half an hour travel time).

(iii) *Closure of* $\mathcal{TTT}$: consider $MM'M'' \in \mathcal{TTT}$, so with $M, M', M'' \in \mathcal{T}$. The case of an internal step, in MM' or M'' is easy, by the closure of $\mathcal{T}$ and $\mathcal{TT}$. The interesting case is again a root step: $MM'M'' \equiv \mathsf{S}ABM'' \rightarrow AM''(BM'')$. Now $A \in \mathcal{T}$, $M'' \in \mathcal{T}$, $B \in \mathcal{T}$, and hence AM'', $BM'' \in \mathcal{T}$, by closure of $\mathcal{TT}$; hence $AM''(BM'') \in \mathcal{TTT}$.

(iv) *Closure of* $\mathcal{T}^{\sim n}$: similar to the reasoning above, but easier because no root steps are possible for $n \geq 4$. We will not need this fact however, and mention it only for completeness' sake.

Note that all reducts of $MM'M'' \in \mathcal{TTT}$ have again the form PQR with $P, Q, R \in \mathcal{T}$. Now we are ready to give 'checkmate'.

Proof (Proof of Theorem 22) Consider $Mxyzu$, for an S-term M. Then $Mx, y, z \in \mathcal{T}$. So $Mxyz \in \mathcal{TTT}$. The only way to consume u is that $Mxyz \twoheadrightarrow \mathsf{S}AB$ for some A, B. But by closure of $\mathcal{TTT}$, $Mxyz$ can only reduce to a term PQR with $P, Q, R \in \mathcal{T}$. Matching $\mathsf{S}AB$ with PQR means that $P \equiv \mathsf{S}$. However, $S \notin \mathcal{T}$.

Remark 26 It might be thought that the fact that S-terms can only consume three variables, follows from the fact that already S itself can only consume 3 variables, has order 3. But more is at stake, as is apparent in a comparison of the order of J-terms, where $\mathsf{J}abcd \rightarrow ab(adc)$. J-terms, built from J by applications, can have arbitrarily large order, as a consideration of terms $\mathsf{J}^n\mathsf{J}$ shows; they have an order growing linearly with their length.

Remark 27 Note that the theorem refers to a string of variables offered as 'food'. If one offers a general sequence of S-terms, an S-term can consume much more than 3 arguments, namely infinitely many arguments. An example is S itself; it actually can consume an unbounded number of S's. The proof is an easy corollary of the fact that a flat S-term is normalizing, together with the observation that an S-term in normal form has the shape S, $\mathsf{S}A$ or $\mathsf{S}AB$.

Remark 28 (Albert Visser, personal communication) A corollary obtained from the order of S-terms is as follows. There is no S-term M such that $Mxy \rightarrow^* M(xy)$, for such an M would have infinite order.

Theorem 29 (Waldmann 2000) S*-terms are top-terminating, that is, there is no reduction containing infinitely many root steps.*

As a consequence, all finite S-terms have the property SN^∞, that is, they are infinitary normalizing (Klop and de Vrijer 2005; Endrullis et al. 2009d). For more intuition and explanation about SN^∞, see Appendix A.5. In the terminology explained in the Appendix, this theorem can be rephrased as:

(i) *There are no mute* S*-terms*, or
(ii) *Every* S*-term has a root normal form (rnf).*

A remark about terminology: instead of *root termination*, Waldmann (2000), Berarducci and Intrigila (1996) use *top termination*.

5.6 S-terms in Lambda Calculus

In the study of λ-calculus and Combinatory Logic much attention has been devoted to a comparison of these two systems with respect to normalization and convertibility properties (Barendregt 1984). We will now do the same with the restriction to S-terms. So we are interested in applicative combinations of the λ-term

$$\mathsf{S} \equiv \lambda abc.\, ac(bc)$$

subject to the β-reduction rule. We have

$$(\lambda abc.\, ac(bc))ABC \to (\lambda bc.\, Ac(bc))BC \to (\lambda c.\, Ac(Bc))C \to AC(BC)$$

Thus rewrite sequences of S-terms in CL can be interpreted directly in λ-calculus. However, reduction in λ-calculus is more fine-grained, thus causing a number of differences. For example, consider the following S-terms from Waldmann (2000):

$$\mathsf{S}(\mathsf{S}T\mathsf{S})A \qquad \text{and} \qquad \mathsf{S}(TA)(\mathsf{S}A\mathsf{S})$$

where $T \equiv \mathsf{SS}$ and $A \equiv \mathsf{SSS}$. In CL, both terms are normal forms and thus not convertible because they are not identical. However, in λ-calculus these terms turn out to have the same β-normal form:

$$\lambda a.aG(aG\lambda b.ab(Gb))$$

where $F \equiv \lambda yb.ab(yb)$ and $G \equiv aF(\lambda b.ab\lambda c.ac(bc))$. So we can conclude that convertibility in CL and λ-calculus differ for S-terms. In both systems the decidability of convertibility is an open problem:

Question 30 Is convertibility of S-terms decidable

(i) in Combinatory Logic?
(ii) in λ-calculus?

For CL this open problem has been mentioned in the literature repeatedly, see for example Scott (1975), Waldmann (1998, 2000), Barendregt (1975).

Question 31 Can Böhm tree techniques help to approach the previous question? It is well known that $M = N$ implies that M and N have the same Böhm tree. The converse does not hold for general λ-terms. However, does it hold for S-terms?

Are there non-convertible (finite) S-terms having equal Böhm trees? If such S-terms do not exist, then convertibility of S-terms can be reduced to equality of their Böhm trees. If these terms exist, then the question can be strengthened to equality of clocked Böhm trees (Endrullis et al. 2010d, 2012b, 2013b)?

It is well-known that convertibility in full CL differs from that in λ-calculus after translation. The difference can however be bridged by a set of 5 axioms for CL known as A_β in Barendregt (1984). Thus we have for CL-terms M, N:

$$\mathrm{CL} + A_\beta \vdash M = N \iff \lambda\beta \vdash \phi(M) = \phi(N)$$

where ϕ is the translation of CL-terms in λ-terms.

Question 32 Can the difference of convertibility of S-terms in CL versus convertibility of their λ-translations analogously be bridged by a finite set of equations between S-terms?

We now consider the preservation of termination when translating S-terms from Combinatory Logic to λ-calculus. Obviously, non-termination is preserved. We will now present an example, showing that this is not the case for termination. We will lead up to this example by a sequence of 'exploding' terms. We call a term *exploding* if its λ-normal form is significantly larger than their normal form in Combinatory Logic. As a first example, consider the term

$$\mathsf{S}TTT\mathsf{SS}\,,$$

from Barendregt et al. (1976, Exercise 7.8), which has 485 symbols in its λ-normal form. The normal form of this term in Combinatory Logic is displayed in Fig. 5.15.

A second, even more exploding term has been found by van der Poel (1975, 1992):

$$\mathsf{S}T\mathsf{S}T\mathsf{S}$$

This term is normalizing in Combinatory Logic as well as λ-calculus. The size of its normal form in λ-calculus consists of 963955 symbols, so a magnitude larger than the size of its normal form in Combinatory Logic

$$\mathsf{S}(\mathsf{S}(TC))(\mathsf{S}(\mathsf{S}(TC))(\mathsf{S}(\mathsf{S}(\mathsf{S}(TC)))(\mathsf{S}(TC)(\mathsf{S}(\mathsf{S}(TC)))))) \quad \text{where } C \equiv \mathsf{S}T\mathsf{S}$$

which consists of 79 symbols (40 S's and 39 application symbols), in other words, the number of nodes in the formation tree of the term.

A third exploding term has been found by Johannes Waldmann:

$$\mathsf{SS}(\mathsf{SSS})(\mathsf{S}(\mathsf{S}(\mathsf{S}(\mathsf{SS}))\mathsf{SSS}))$$

The normal form in Combinatory Logic has 25 symbols and has the astronomical size

$$1148661513459923354761019888458808398999497$$

in λ-calculus (approximately 10^{42}).

As a climax to this crescendo, we have the S-term in the following example.

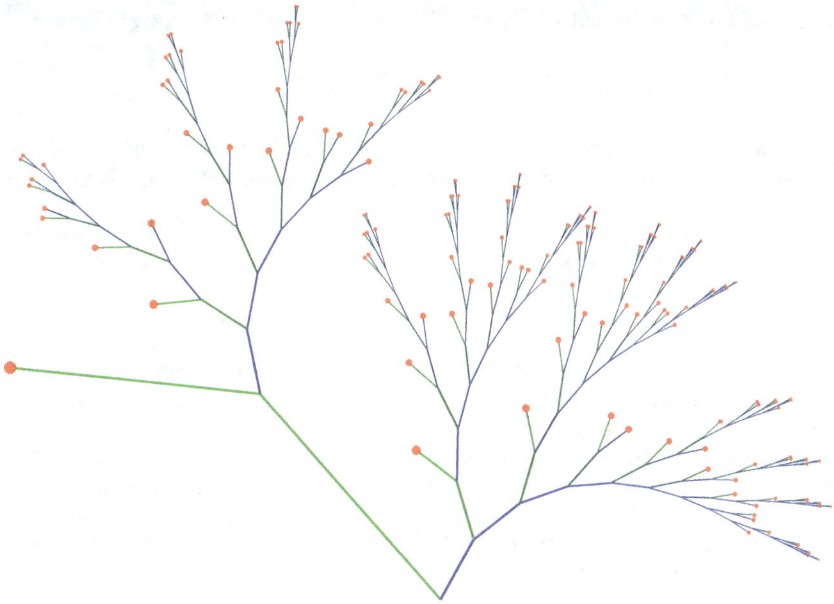

Fig. 5.15 Normal form of S(SS)(SS)(SS)SS. The green left spine has length 2, the blue right spine has length 9

Example 33 (Waldmann 2000). The S-term

$$\mathsf{S}(\mathsf{S}(\mathsf{S}TT))(\mathsf{S}(\mathsf{S}TT))$$

is in normal form, but its translation to λ-calculus admits an infinite reduction. To see this we use the following abbreviations:

$$T \equiv \mathsf{SS} \qquad\qquad Z \equiv \mathsf{S}TT \qquad\qquad V \equiv \mathsf{S}Z$$

Then $\mathsf{S}(\mathsf{S}(\mathsf{S}TT))(\mathsf{S}(\mathsf{S}TT)) \equiv \mathsf{S}VV \rightarrow \lambda z.\, Vz(Vz)$. For $n \in \mathbb{N}$ we have

$$\begin{aligned} Vz(T^n(Vz)) &\equiv \mathsf{S}Zz(T^n(Vz)) \rightarrow \underline{Z\,(T^n(Vz))}\,(z(T^n(Vz))) \\ &\triangleright Z\,(T^n(Vz)) \equiv \mathsf{S}TT(T^n(Vz)) \rightarrow T^{n+1}(Vz)\;(T^{n+1}(Vz))\,. \end{aligned}$$

Moreover, for $m, n \in \mathbb{N}$ we have

$$\begin{aligned} T^{m+1}(Vz)(T^n(Vz)) &\equiv \mathsf{SS}(T^m(Vz))\;(T^n(Vz)) \\ &\rightarrow \mathsf{S}(T^n(Vz))\;\underline{(T^m(Vz)\;(T^n(Vz)))} \\ &\triangleright T^m(Vz)\;(T^n(Vz)) \end{aligned}$$

By induction on m, we get $T^{m+1}(Vz)(T^n(Vz)) \; (\to \cup \rhd)^* \; Vz(T^n(Vz))$. Hence

$$Vz(T^n(Vz)) \quad (\to \cup \rhd)^* \quad Vz(T^{n+1}(Vz))$$

for all $n \in \mathbb{N}$, giving rise to an infinite rewrite sequence. Note that the patterns appearing in this infinite reduction are reminiscent of the ones in the Zachos reduction.

We have seen that termination of S-terms in Combinatory Logic is decidable (Theorem 9). However, this problem remains open for S-terms in λ-calculus.

Question 34 Is termination of S-terms in λ-calculus decidable?

Experiments suggest that the set of terminating S-terms in λ-calculus is not regular. Note that by contrast, this set is regular in Combinatory Logic (Theorem 9).

Acknowledgements We thank Hans Zantema for stimulating discussions about S-terms during recent years, and in particular for recently suggesting the spiralling graphical rendering of S-terms as in our figures. We also thank Jan Bergstra, Vincent van Oostrom, Wim van der Poel, Richard Statman, and Albert Visser for contributing their insights. We also thank the referees for many helpful corrections, comments and suggestions.

A A Refresher of Finitary and Infinitary Rewriting

A.1 Introduction

We will presuppose a modest familiarity with Combinatory Logic and λ-calculus, but even so a quick refresher of the main notions and notations, tailored towards the present walk in Smullyan's forest, may be useful. The experienced reader, familiar with Barendregt (1984) or Terese (2003) will have little need for these preliminaries, and can quickly glance over the next pages.

A.2 Finitary Combinatory Logic

Combinatory Logic (CL) is based on the basic combinators S, K, I with reduction rules

$$\begin{aligned} \mathsf{S}xyz &\to xz(yz) \\ \mathsf{K}xy &\to x \\ \mathsf{I}x &\to x \end{aligned}$$

Terms are formed from S, K, I using application, written sometimes as @ but in our paper as · (in the figures as •). The · is usually not written in the term formation, but visible in figures where terms are depicted as labelled trees.

We assume that the reader is familiar with term formation, with its usual bracketing convention called *association to the left*. Let us stress here that term formation here is governed by induction; later on we will invoke also coinduction to generate possibly infinite terms. The set of terms of CL will be denoted as *Ter*(CL). Closed terms are just built from S, K, I; open terms may contain variables $x, y, z, a, b, c, \ldots$. Closed CL-terms are also called *combinators*. We employ $\rightarrow$ for one step reduction, which may be performed in a context, a term with a hole in it. E.g. S□S is a context with one hole. We often write $C[\ ]$ for a context, and write $C[M]$ for the result of substituting M for the holes in $C[\ ]$.

We use $\twoheadrightarrow$ or $\rightarrow^*$ for the transitive reflexive closure of $\rightarrow$. Another reminder: a *redex* is an instance of a left-hand side of one of the reduction rules, and a *normal form* is a term without a redex occurrence.

We use '$=$' for the convertibility relation defined by the reduction rules; that is the equivalence generated by $\rightarrow$. It should be distinguished from '$\equiv$', which denotes syntactical equality. We will also use $\equiv$ for abbreviations ('definitional equality'), and write e.g. $A \equiv \mathsf{SSS}$, where A abbreviates SSS. Sometimes we write $M =_{\mathsf{I,K,S}} N$ or $M =_{\mathsf{S}} N$ if we want to indicate the rules generating the convertibility relation. Later on we will also use $\equiv$ for syntactical equality of infinite terms. Note that when we use $\equiv$ for finite or infinite λ-terms, we work modulo α-conversion. So $\lambda x.\, x \equiv \lambda y.\, y$, and to give a non-trivial example for infinite λ-terms:

$$\lambda x_0 x_1 x_2 \ldots x_n \ldots \equiv \lambda x_{\pi(0)} x_{\pi(1)} x_{\pi(2)} \ldots x_{\pi(n)} \ldots$$

for any permutation π of the natural numbers.

S-terms, CL(S)

In this paper we will mostly be concerned with a 'subsystem' of full CL, to be called CL(S), where terms are only generated from the basic combinator S, excluding I and K, and including possibly variables, by means of the binary application. *Ter*(CL(S)) is the set of S-terms obtained this way. Thus CL(S) with the S-reduction rule is just as CL, an example of an *orthogonal* TRS, term rewriting system. Hence it satisfies the confluence or Church-Rosser property: CL(S) $\models$ CR.

S-terms are Non-erasing

In contrast with full CL, the sub-TRS CL(S) satisfies also the property NE, *non-erasingness*: CL(S) $\models$ NE. This means that there is no erasure of subterms, because the S-rule has the same variables in its left-hand side and right-hand side. This has an important consequence, namely that for S-terms the properties of WN (Weak Normalization, having a normal form) and SN (Strong Normalization, there are no infinite reductions, so eventually every reduction must end in a normal form) are equivalent. For λ-calculus there is an analogous situation, in that for λI-calculus the Theorem of Church holds, stating that for a λI-term the properties WN and SN coincide.

A.3 *Finitary λ-calculus*

Again, we will not dwell on the very first notions of λ-calculus, such as term formation, free and bound variables, the β-reduction rule and so on. These can be found in any introduction. As to the notation, we use this similarly to that of CL, so $\rightarrow$, $=$, $\twoheadrightarrow$, $\equiv$ have meanings analogous to those for CL, except that they now refer to the β-reduction rule. We will in this paper not consider the η-reduction rule. If clarity is served, and we are working in λ-calculus, we will sometimes subscript the reduction symbols with β, so $\rightarrow_\beta$, $=_\beta$, $\twoheadrightarrow_\beta$. As before, a normal form is a term without a redex occurrence, in this case without β-redex.

There is an important subcalculus of λ-calculus, called λI-calculus, which we just mentioned. Here in an abstraction term $\lambda x.\, M$, the variable x must occur free in M. So e.g. $\lambda xy.\, x$ is not a λI-term. Abbreviating these calculi by just λ and λI, we have $\lambda \models \mathrm{CR}$, also $\lambda\mathrm{I} \models \mathrm{CR}$, and

$$\lambda\mathrm{I} \models \mathrm{WN}(M) \Leftrightarrow \mathrm{SN}(M)$$

for all $M \in \mathit{Ter}(\lambda\mathrm{I})$; this last assertion is known as Church's Theorem.

The Order of Terms in CL and λ Calculus

In applicative rewrite systems such as CL and λ-calculus, there is an intuition of arguments being 'consumed' or 'eaten' by the term of which they are arguments. This gives rise to the notion of the *order* of a term. In the words of Curry and Feys (1974): *Roughly speaking, the order of an ob X is the number of variables in the combination* $Xx_1x_2\ldots x_n$ *which are acted on by* X, *or the minimum number of arguments to X considered as a function.*

For λ-calculus, there are two equivalent definitions, with one of them also pertaining to CL.

Definition 35 Let $M \in \mathit{Ter}(\lambda)$.

(i) M is of *order* 0 (also: M has order 0, or: *M is a 0-term*) if there is no term P such that $M \twoheadrightarrow_\beta \lambda x.\, P$.

(ii) M is of order $n+1$ if $M \twoheadrightarrow_\beta \lambda x.\, P$ where P is of order n.
(iii) M is of order ∞ (has infinite order), if M does not have a finite order.

Example 36 (i) Variables have order 0; $\Omega \equiv \omega\omega \equiv (\lambda x.\, xx)(\lambda x.\, xx)$ has order 0. The term $Y\mathsf{K}$, where Y is any fixed point combinator and $\mathsf{K} \equiv \lambda xy.\, x$, has infinite order. It is called sometimes the Ogre (Omnivore) or K^∞.

(ii) If M is of order 0, so is MP for any P.
(iii) If M is of order ∞, so is MP for any P.
(iv) If M is of order $n+1$, Mx is of order n. (Avoid the pitfall: if M is of order n, MN can have a higher order than M.)
(v) (Berarducci 1996) If $M =_\beta MP$ for some P, M is of order 0.
(vi) The order is invariant under $=_\beta$, β-conversion.
(vii) (Curry and Feys 1974, p. 148, Theorem 5). If M has a normal form, then M has a finite order.
(viii) If M has a head normal form, then M has a finite order.
(ix) (Curry and Feys 1974, p. 148, Theorem 6). If M is of order $\geq m$, and $x_1, \ldots x_m$ are not free in M, then

$$\lambda x_1, \ldots x_m.\, M x_1, \ldots x_m =_\beta M \ .$$

(x) To be read after introduction of infinitary λ-calculus, and Berarducci Trees $\mathsf{BeT}(M)$: If M has infinite order, then M has the infinite normal form

$$\lambda x_0 x_1 x_2 \ldots x_n \ldots \equiv \mathsf{BeT}(M)$$

(Note that there is only one such infinite λ-abstraction, due to α-conversion.)

Here is an alternative and equivalent definition of the order of a term, this time also applying to CL. It is from Curry and Feys (1974, p. 144).

Definition 37 Let $M \in Ter(\lambda)$, or let $M \in Ter(\mathrm{CL})$. The *order* of M is the least m such that every β-reduction of

$$M x_1 \ldots x_m y_1 \ldots y_n$$

is obtained by applications of rule (ν) to a β-reduction of

$$M x_1 \ldots x_m$$

Here rule (ν) (see Curry and Feys 1974, p. 59) is the *left-monotony rule* used in defining the reduction relation $\twoheadrightarrow_\beta$ or $\twoheadrightarrow_{\mathrm{CL}}$, as follows:

$$M \twoheadrightarrow N \implies MZ \twoheadrightarrow NZ$$

So, the string of variables $\boldsymbol{y}$ plays a dummy role in the reduction of $M\boldsymbol{x}\boldsymbol{y}$.

The notion of order, in particular order 0, is also involved in the notion of 'undefined', or 'meaningless'. See also Barendregt (1984), on representing 'undefined' in the λ-calculus.

We summarize the various notions of 'undefined' appearing in this paper, as follows. The implications as displayed are the only ones valid for these notions, of course apart from those obtained by transitivity of implication. For an axiomatic theory about the notion of 'undefined', we refer to Terese (2003), and in particular to Kennaway et al. (1999).

Theorem 38 *The following notions of 'undefined' Kennaway et al. (1999), Barendregt (1992a) are strictly ordered by implication:*

$$\begin{array}{c}
\text{no root normal form, (aka mute, root active)} \\
\Downarrow (i) \\
\text{no weak head normal form; not for CL} \\
\Downarrow (ii) \\
\text{order } 0 \\
\Downarrow (iii) \\
\text{no head normal form, (aka unsolvable)} \\
\Downarrow (iv) \\
\text{no normal form, (aka not weakly normalizing)} \\
\Downarrow (v) \ (\text{for } \lambda\text{I} - \text{terms, for } \mathsf{S} - \text{terms } \Updownarrow) \\
\text{non} - \text{terminating, (aka not strongly normalizing)}
\end{array}$$

Proof (i) Note that a weak head normal form (whnf) has the form $xM_1 \ldots M_n$ or $\lambda x.M$. Both forms of whnf are also root normal forms (rnf).

(ii) Suppose M is not of order 0. Then M reduces to an abstraction. But that is a whnf. The reverse does not hold: if M is a zero term, it does not follow that M has no whnf; e.g. $x\boldsymbol{P}$ is of order 0, but also a whnf.

(iii) Suppose M is of order 0. Then it cannot consume any argument. But then it cannot be 'solved towards' I, i.e. there is no $\boldsymbol{P}$ such that $M\boldsymbol{P} \twoheadrightarrow_\beta \mathsf{I}$. A theorem by Wadsworth (Barendregt 1984, Theorem 8.3.14, p. 175) states that therefore M has no hnf.

(iv) Implications (iv) and (v) are standard theory.

Remark 39 Theorem 38 shows that the notion of undefinedness embodied by mute terms is the strongest one of those featuring in this paper. That mute is 'more undefined' than unsolvable, apart from the strict inclusion mute $\Longrightarrow$ unsolvable, is witnessed by a theorem of Berarducci (1996), stating that it is consistent to put all mute terms uniformly equal to an arbitrary λ-term Q. This is not the case for unsolvable terms. Even the single equation $\Omega_3 = \mathsf{I}$ generates inconsistency when added to λ-calculus Jacopini (1975). It is a simple exercise to find the few deduction steps showing $x = y$ for different variables x and y.

Translating CL in λ

There are translations in both directions between CL and λ, preserving certain properties. Here we will only be concerned with the easy direction, from CL to λ. The translation consists of the contextual closure of the replacements

$$\begin{aligned} \mathsf{S} &\mapsto \lambda xyz.\,xz(yz) \\ \mathsf{K} &\mapsto \lambda xy.\,x \\ \mathsf{I} &\mapsto \lambda x.\,x \end{aligned}$$

To keep reading light, we allow ourselves the notational liberty of using the same letter for a term in CL and its translation in λ. The reader who dislikes that, will have no problem in inserting several $(\ldots)_\lambda$ as used in Barendregt (1984) at the right places. The salient property is that reduction is preserved:

$$\mathrm{CL} \models M \twoheadrightarrow N \implies \lambda \models M \twoheadrightarrow_\beta N$$

and restricted to S-terms:

$$\mathrm{CL}(\mathsf{S}) \models M \twoheadrightarrow N \implies \lambda\mathrm{I} \models M \twoheadrightarrow_\beta N$$

Of course this translation raises the question whether the properties WN and SN are also preserved. For full CL, it is well-known that WN and SN are not preserved. For instance, $M \equiv \mathsf{S}(\mathsf{K}\omega)(\mathsf{K}\omega)$ where $\omega \equiv \mathsf{SII}$, is a CL-normal form, but its translation into λ has no β-normal form, as can be easily verified.

It could be thought that this phenomenon is due to the erasing nature of K, but similar counterexamples exist in the non-erasing version of CL, called CL_I, which has atomic combinators I, S, B, C with for B, C the rules

$$\mathsf{B}xyz \to x(yz) \qquad \text{and} \qquad \mathsf{C}xyz \to xzy$$

Then we have counterexamples $\mathsf{B}(\mathsf{CI}\omega)(\mathsf{C}\omega)$ and $\mathsf{S}(\mathsf{CI}(\mathsf{CI}\omega))(\mathsf{C}\omega)$. Remarkably, just with S alone there is already such a counterexample, as we will see in the paper.

From the preservation of reduction we immediately have the preservation of convertibility:

$$\mathrm{CL} \models M =_{\mathsf{I,K,S}} N \implies \lambda \models M =_\beta N$$

and restricted to S-terms:

$$\mathrm{CL}(\mathsf{S}) \models M =_\mathsf{S} N \implies \lambda\mathrm{I} \models M =_\beta N$$

Also here it is well-known (Barendregt 1984, p. 157) that the reverse does not hold. The classical example is the equation $\mathsf{SK} = \mathsf{KI}$: $\mathrm{CL} \not\models \mathsf{SK} = \mathsf{KI}$, but $\lambda \models \mathsf{SK} =_\beta \mathsf{KI}$.

For the restriction to S-terms we have the same non-preservation, but more complicated; abbreviating $T \equiv \mathsf{SS}$ and $A \equiv \mathsf{SSS}$ we have, as elaborated in the paper: $\mathsf{CL} \nvDash \mathsf{S}(\mathsf{S}T\mathsf{S})A = \mathsf{S}(TA)(\mathsf{S}A\mathsf{S})$, but $\lambda \models \mathsf{S}(\mathsf{S}T\mathsf{S})A =_\beta \mathsf{S}(TA)(\mathsf{S}A\mathsf{S})$.

It is an interesting question whether the order of a term is preserved under translation from CL to λ-calculus; indeed we expect so.

Conjecture 40 Let $M \in \mathit{Ter}(\mathrm{CL})$. Then $\mathsf{order}(M) = \mathsf{order}((M)_\lambda)$.

Example 41 (i) The CL-term SS has order 3, because $\mathsf{SS}xyz \twoheadrightarrow_{\mathrm{CL}} yz(xyz)$, and the last term is of order 0. Also $(\mathsf{SS})_\lambda \twoheadrightarrow_\beta \lambda xyz.\, yz(xyz)$, again of order 0, but now in λ-calculus.

(ii) The CL-term SSS has order 2, because $\mathsf{SS}xy \twoheadrightarrow_{\mathrm{CL}} xy(\mathsf{S}xy)$, a term of order 0. Also $\mathsf{S}(\mathsf{SS})_\lambda \twoheadrightarrow_\beta \lambda ab.\, ab(\lambda c.\, ac(bc))$, of order 2.

(iii) SSSS has order 2, both in CL and λ-calculus.

(iv) Let $\omega \equiv \mathsf{SII}$. Then $\mathsf{S}(\mathsf{K}\omega)(\mathsf{K}\omega)$ has order 1, both in CL and λ-calculus.

(v) S(SS)(SSS) has order 2, both in CL and λ-calculus.

Conjecture 42 Let $M \in \mathit{Ter}(\mathrm{CL}(\mathsf{S}))$. Then $Mxyz$ normalizes $\Longleftrightarrow$ $(M)_\lambda$ normalizes.

Conjecture 43 Let $M \in \mathit{Ter}(\mathrm{CL})$. Then $Mx_1x_2 \ldots x_{\mathsf{order}(M)}$ normalizes $\Longleftrightarrow$ $(M)_\lambda$ normalizes.

A.4 Head Reduction and Root Reduction

The evaluation of λ-terms to their Böhm Trees (BT's) proceeds from left to right, starting with the head. In this section of the Appendix we will present definitions and the main facts about head reduction. But there is a complementing point of view, where evaluation proceeds top-down, from the root to lower parts of the term. That leads to root reduction, and to a finer kind of semantics, embodied in so-called Berarducci Trees (BeT's). Here 'finer' refers to the semantics as given by Böhm Trees. We will explain this in more detail below. BT's have more computational significance, whereas BeT's are a beautiful tool in consistency studies of λ-calculus Berarducci (1996), Berarducci and Intrigila (1996).

Head and Root Reduction in CL

CL-terms can be conveniently drawn as finite term trees, with binary application nodes, and S's as terminal nodes; see the reduction of AAA where $A \equiv \mathsf{SSS}$ in Fig. 5.2. The (left) spine of a CL-term M is the leftmost branch in the term tree of M. The spine length $\mathsf{sl}(M)$ of M is the length of the spine, defined as the number of steps on the spine. So $\mathsf{sl}(\mathsf{S}) = 0$, $\mathsf{sl}(\mathsf{S}X) = 1$, $\mathsf{sl}(\mathsf{S}XY) = 2$, $\mathsf{sl}(\mathsf{S}XYZ) = 3$ and so on. Here X, Y, Z are metavariables for CL-terms. Note that an S-term is a redex iff its spine has length 3.

A *head normal form* (hnf) in CL is a term starting with I, K, or S, but without enough arguments to constitute a redex; so I, K, S themselves, and also terms of the form $\mathsf{K}X$, $\mathsf{S}X$, $\mathsf{S}XY$. So restricting to terms only containing S's, an S-term is a hnf if its spine length is 0, 1 or 2.

A redex R in M is called the *head* redex if it is in head position, i.e. $M \equiv RN_1N_2 \dots N_k$, for a sequence of $k \geq 0$ arguments. If $k = 0$, the redex is called the *root* redex. If a term is not a hnf, it has a unique head redex. A root redex is also a head redex, but not vice versa. See Fig. 5.2. Note that the head redex if it exists, is located at the end of the spine, containing in the case of S-terms the last three steps of the spine.

We can give an equivalent alternative definition of a hnf: this is a term that cannot reduce to a term having a head redex. This definition leads us to the more complicated notion of a *root normal form* (rnf); this is a term which cannot reduce to a term which has (or equivalently, is) a root redex (see also Middeldorp 1997). So hnf's are a fortiori rnf's, because a root redex is also a head redex. Restricting to S-terms: an S-term is a rnf if it is of the form S, $\mathsf{S}X$, $\mathsf{S}XY$, or XY where X cannot reduce to a term of the form $\mathsf{S}PQ$. Indeed in all these 4 cases the root node of the term tree can never be active in a reduction. The complication of the notion rnf is its undecidability; by contrast: the property of being a hnf is trivially (syntactically) decidable. Indeed for full CL, this property is undecidable. But remarkably, for the subsystem of CL(S), the notion of rnf is decidable![4]

Example 44 The term $AA\mathsf{AS}$ is an example of a S-term which is in root normal form since the root-application never can be 'touched'. The term is not in head normal, it does not even reduce to a head normal form, because AAA is of order 0.

Head and Root Reduction in λ-calculus

Remember that every λ-term M is of one of the following forms:

$$M \equiv \lambda x_1 \dots x_n.\, yM_1 \dots M_m$$

or

$$\lambda x_1 \dots x_n.\, (\lambda y.\, P)QM_1 \dots M_k$$

In the first case M is said to be a *head normal form* (hnf); in the second case M has the *head redex* $(\lambda y.\, P)Q$.

[4]Note that S-term M is a rnf $\Leftrightarrow$ M is of the form x, S, $\mathsf{S}A$, $\mathsf{S}AB$ or $M_1\, M_2$ with M_1 of order 0, i.e. M_1 has an infinite head reduction, i.e. M_1 is accepted by the automaton in Fig. 5.14. So 'M is rnf' is decidable with the proviso that the automaton in Fig. 5.14 has yet to be proved correct.

Head and Root Normalization Theorems

We are approaching the fundamental notions of Böhm Trees and Berarducci Trees for CL and λ-calculus. Before we are there, there are two important theorems, one for 'head' and one for 'root', that must be presented.

Theorem 45 *The following are equivalent, for CL and λ-calculus:*

(i) M has no hnf.
(ii) M has an infinite reduction of consecutive head steps:

$$M \equiv M_0 \rightarrow_h M_1 \rightarrow_h M_2 \rightarrow_h \ldots$$

(iii) M has an infinite reduction containing infinitely many head steps.

We will now turn to root reduction instead of head reduction. Let us first define the notion of root redex and root normal form (rnf).

Definition 46 (i) Let R be a redex in M. Then R is the root redex of M, if $M \equiv R$.
(ii) M is a *root normal form* (rnf), also called root stable, if M cannot reduce to a redex.
(iii) M has a rnf, if it reduces to a rnf.

There is a considerable analogy between these two notions, hnf and rnf, but not all the way. To appreciate the intricacy of the notion rnf, consider the statement 'M has no rnf'. (Also called: M is root-active.) This means: M does not reduce to a term which cannot reduce to a redex. In a more understandable reformulation: *every reduct of M can further reduce to a redex.*

Theorem 47 *M has no root normal form iff M admits a reduction with infinitely many root steps.*

For completeness sake, we mention the following. Note that for orthogonal systems, SN^∞ is the same as root normalization; see Theorem 29 and the subsequent discussion.

Theorem 48 (Ketema et al. 2005) *For orthogonal TRSs, CL, CL(S) it holds that:* $SN^\infty \implies$ *there are no reduction cycles.*

Corollary 49 S-*terms are acyclic.*

A.5 Infinitary Rewriting

We can easily introduce infinite terms in CL and λ-calculus by interpreting the formerly inductive term formation rules in a coinductive way. This is by now a standard definition mechanism that needs no further explanation here; see e.g. Endrullis and Polonsky (2013), Endrullis et al. (2013a).

However, we can also introduce infinite terms by the even more classical mechanism of metric completion, as below.

As to reduction (rewriting) of infinite terms we observe that the rules apply just as well to finite as to infinite terms; their applicability just depends on the presence of a finite 'redex pattern'. In the following, the signature Γ stands for either λ or CL. Infinite terms arise from the set of finite terms, $Ter(\Gamma)$, by metric completion, using the well-known distance function d such that for terms t, s, $d(t, s) = 2^{-n}$ if the n-th level of the terms t, s (viewed as labelled trees) is the first level where a difference appears, in case t and s are not identical; furthermore, $d(t, t) = 0$. It is standard that this construction yields $\langle Ter(\Gamma), d\rangle$ as a metric space. Now infinite terms are obtained by taking the completion of this metric space, and they are represented by infinite trees. We will refer to the complete metric space arising in this way as $\langle Ter^{\infty}(\Gamma), d\rangle$, where $Ter^{\infty}(\Gamma)$ is the set of finite and infinite terms over Γ.

A natural consequence of this construction is the emergence of the notion of *Cauchy convergence*: we say that $t_0 \to t_1 \to t_2 \to \dots$ is an infinite reduction sequence with limit t, if t is the limit of the sequence $t_0, t_1, t_2, \dots$ in the usual sense of Cauchy convergence. In fact, we will use throughout a stronger notion that has better properties. This is *strong convergence*, which in addition to the stipulation for Cauchy convergence, requires that the depth of the redexes contracted in the successive steps tends to infinity. So this rules out the possibility that the action of redex contraction stays confined at the top, or stagnates at some finite level of depth.

The notion of *normal form*, which now may be an infinite term, is unproblematic: it is a term without a redex occurrence.

The most important properties of finite and infinitary rewriting are shown in Table 5.1.

Here the left column states the finitary properties, and the right column states the analogous properties for the infinitary case.

Most of the properties in this table are self-explanatory, but SN^{∞} requires some special explanation. The helpful underlying analogy is stated in the first two rows of the table: infinite reductions in the finite realm (not delivering a resulting finite normal form) are analogous to 'not strongly convergent', i.e. divergent reductions in the infinitary realm of rewriting. Here not strongly convergent, divergent, means that there is stagnation at some problematic point in the term being rewritten; the depth of the contracted redex is then not increasing as required for strong convergence. This means that the root of the subterm-tree at that problematic point refuses to 'stabilize'. Now SN^{∞} states the absence of such divergences, non-stabilizations. Thus we see that SN^{∞} is tantamount to 'root stabilization', uniformly for all roots of subterms that are encountered.

In our paper we have used sometimes as synonym for SN^{∞} the phrase 'root stabilization'. In the terminology of Berarducci Trees (BeT's) this means that no mute terms occur, or, that the BeT is $\perp$-free, total. This is the exceptional situation for finite S-terms, they are SN^{∞}, infinitary normalizing. Not so for general CL-terms, and also not for infinite S-terms.

Table 5.1 The main properties in finite and infinitary rewriting

Finitary rewriting	**Infinitary rewriting**
Finite reduction	Strongly convergent reduction
Infinite reduction	Divergent reduction ("stagnating")
Normal form	(Possibly infinite) normal form
CR: two coinitial finite reductions can be prolonged to a common term	CR^{∞}: two coinitial strongly convergent reductions can be prolonged by strongly convergent reductions to a common term
UN: two coinitial reductions ending in normals forms, end in the same normal form	UN^{∞}: two coinitial strongly convergent reductions ending in (possibly infinite) normal forms, end in the same normal form
SN: all reductions lead eventually to a normal form	SN^{∞}: all reductions lead eventually to a (possibly infinite) normal form, equivalently: there is no divergent reduction
WN: there is a finite reduction to a normal form	WN^{∞}: there is a strongly convergent reduction to a (possibly infinite) normal form

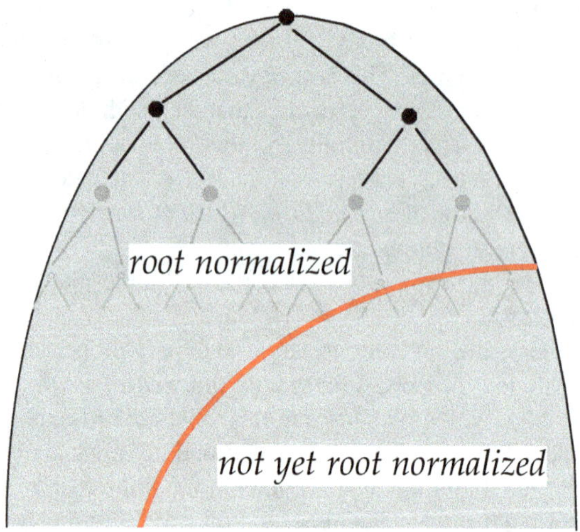

Fig. 5.16 Partially root normalized infinite term tree

That S-terms are SN^{∞} 'infinitary normalizing' means also that whatever reduction we perform, eventually we will reach the unique, possibly infinite normal form. That is, if we apply a fair rewrite strategy, and do not omit a redex forever in our evaluation. This means that even a random infinite reduction of a finite S-term will have a limit, as a partially evaluated tree, one could say, a partially evaluated Berarducci Tree. In our paper we have encountered this in Fig. 5.5, where we have AAA partially evaluated by rewriting the head redex whenever it is present; in the limit there is no head redex anymore – but plenty of other redexes, waiting for their turn. Figure 5.16 displays the general situation of such a partially evaluated infinite term tree.

A.6 *Infinite Normal Forms, Böhm Trees and Berarducci Trees*

Infinitary lambda-calculus, λ_β^∞-calculus and infinitary CL, CL^∞, are the framework for several notions of infinite normal forms that need to be carefully distinguished. We will give an informal exposition.

(i) The easiest notion is that of *infinitary normal forms* as elements of λ_β^∞-calculus or CL^∞. These are just possibly infinite (term) trees without β-redex, or in the case of CL, without I-,K-,S-redex.

(ii) *Böhm Trees* (*BT's*). The introduction of BT's requires a signature extension, with a constant symbol $\perp$ for undefined terms. The essence of BT's is that we evaluate a λ-term ad infinitum by recording its head normal forms, as they present themselves in our branching evaluation, see the definition below. We use $\perp$ where head normalization is not possible. To know that, we need an oracle, as it is not decidable. The well-known definition is:

$$\mathsf{BT}(M) = \begin{cases} \lambda \boldsymbol{x}.\, y\, \mathsf{BT}(M_1) \ldots \mathsf{BT}(M_m) & \text{if } M \text{has hnf} \lambda \boldsymbol{x}.\, y M_1 \ldots M_m, \\ \perp & \text{otherwise.} \end{cases}$$

(iii) *BerarducciTrees* (*BeT's*). Also here we need $\perp$, now to designate subterms that have no root normal form, also known as *mute* or *root-active* terms. The well-known analogous definition is:

$$\mathsf{BeT}(M) = \begin{cases} y & \text{if } M \twoheadrightarrow y, \\ \lambda x.\, \mathsf{BeT}(N) & \text{if} M \twoheadrightarrow \lambda x.\, N, \\ \mathsf{BeT}(M_1)\, \mathsf{BeT}(M_2) & \text{if } M \twoheadrightarrow M_1\, M_2 \text{such that } M_1 \text{is of order0} \\ \perp & \text{in all other cases (i.e., when} M \text{is mute.)} \end{cases}$$

BeT's give finer information than BT's, in the sense that $\mathsf{BT}(M) \leq \mathsf{BeT}(M)$ where the ordering $\leq$ refers to refinement of $\perp$'s; in other words, $\mathsf{BT}(M)$ can be obtained from $\mathsf{BeT}(M)$ by chopping off some subtrees and replacing them by $\perp$. An extreme example is AAA: $\mathsf{BT}(AAA) \equiv \perp \leq \mathsf{BeT}(AAA) \equiv$ the tree in Figs. 5.6 and 5.7. Note that if $\mathsf{BT}(M)$ is $\perp$-free, $\mathsf{BT}(M) \equiv \mathsf{BeT}(M) \equiv$ the normal form of M in λ_β^∞-calculus. An example is the Zachos term in Fig. 5.10. The same is true for BeT's; when they are $\perp$-free, they coincide with the normal form of M in λ_β^∞-calculus. An example is $\mathsf{BeT}(\Omega_3 \mathsf{I})$, as in Fig. 5.18. Some more examples of BT's and BeT's are given in Fig. 5.17.

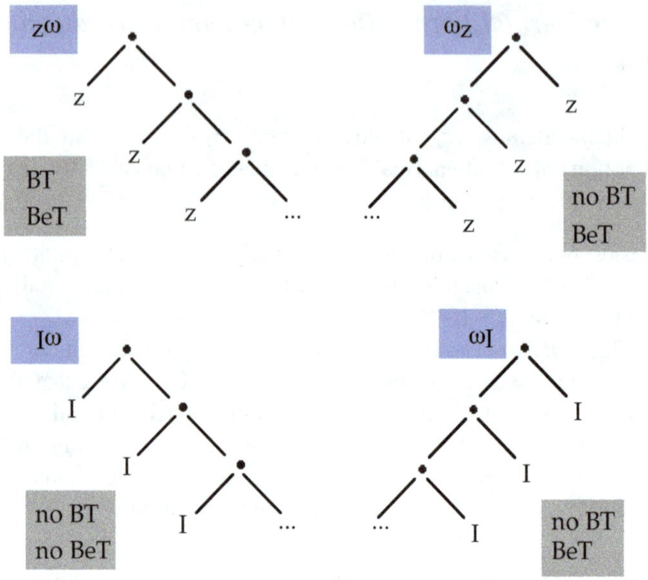

Fig. 5.17 Four infinite λ-terms with their tree families identified, see further Bethke et al. (2000)

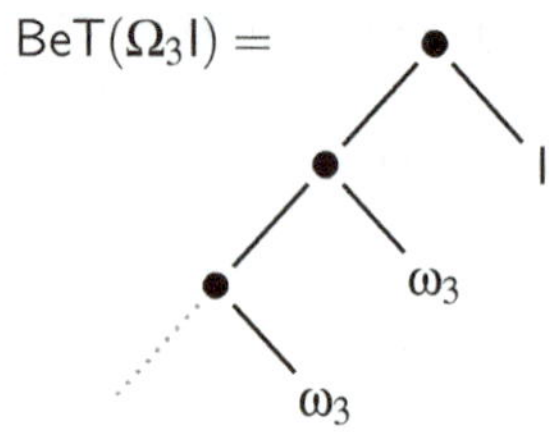

Fig. 5.18 The Berarducci Tree of $\Omega_3 I$, where $I \equiv \lambda x.x$ and $\Omega_3 \equiv \omega_3\omega_3 \equiv (\lambda x.xxx)(\lambda x.xxx)$. It is a Berarducci Tree (BeT), but not a Böhm Tree (BT), as it has an infinite left spine

References

Baader, F., & Nipkow, T. (1998). *Term rewriting and all that*. Cambridge: Cambridge University Press.

Barendregt, H. (1971). *Some extensional term models for combinatory logics and λ-calculi*. Ph.D. thesis, University Utrecht.

Barendregt, H. (1975). Problem 97 of the RTA List of Open Problem. http://www.win.tue.nl/rtaloop/problems/97.html.

Barendregt, H. (1984). *The lambda calculus, its syntax and semantics*, Vol. 103 of *Studies in Logic and The Foundations of Mathematics*. North-Holland, revised edition.

Barendregt, H. (1992a). Representing 'undefined' in lambda calculus. *Journal of Functional Programming*, *2*(3), 367–374.

Barendregt, H. (1992b). Lambda calculi with types. In S. Abramsky, D. M. Gabbay, & T. S. E. Maibaum (Eds.), *Handbook of logic in computer science* (Vol. 2, pp. 117–309). Oxford: Oxford University Press.

Barendregt, H., Bergstra, J., Klop, J. W., & Volken, H. (1976). Degrees, reductions and representability in the lambda calculus. Technical report, Department of Mathematics, Utrecht University, 1976. The report is available via the website http://repository.ubn.ru.nl/handle/2066/20381.

Barendregt, H., Kennaway, R., Klop, J. W., & Sleep, M. R. (1987). Needed reduction and spine strategies for the lambda calculus. *Information and Computation, 75*(3), 191–231.

Barendregt, H., & Klop, J. W. (2009). Applications of infinitary lambda calculus. *Information of Computation, 207*(5), 559–582.

Berarducci, A. (1996). Infinite λ-calculus and non-sensible models. *Logic and algebra (Pontignano, 1994)* (pp. 339–377). New York: Dekker.

Berarducci, A., & Intrigila, B. (1996). Church–Rosser λ-theories, infinite λ-calculus and consistency problems. In *Logic: From foundations to applications* (pp. 33–58).

Bethke, I., Klop, J. W., & de Vrijer, R. C. (2000). Descendants and origins in term rewriting. *Information and Computation, 159*(1–2), 59–124.

Bimbó, K. (2011). *Combinatory logic: Pure, applied and typed*. London: Chapman and Hall.

Cardone, F., & Hindley, R. (2006). History of lambda-calculus and combinatory logic. In D. M. Gabbay & J. Woods (Eds.), *Handbook of the history of logic* (Vol. 5). Elsevier.

Church, A. (1941). *The calculi of lambda conversion*. Vol. 6 of *Annals of Mathematics Studies*. Princeton: Princeton University Press.

Curry, H. B., & Feys, R. (1974). *Combinatory logic* (Vol. I). Amsterdam: North-Holland.

Curry, H. B., Hindley, R., & Seldin, J. P. (1972). *Combinatory logic* (Vol. II). Amsterdam: North-Holland.

Dershowitz, N., Kaplan, S., & Plaisted, D. A. (1991). Rewrite, rewrite, rewrite, rewrite, rewrite. *Theoretical Computer Science, 83*(1), 71–96.

Endrullis, J., de Vrijer, R., & Waldmann, J. (2009a). Local termination. In *Proceedings Conference on Rewriting Techniques and Applications (RTA 2009)*, Vol. 5595 of *LNCS* (pp. 270–284). Berlin: Springer.

Endrullis, J., de Vrijer, R. C., & Waldmann, J. (2010a). Local termination: Theory and practice. *Logical Methods in Computer Science, 6*(3).

Endrullis, J., Geuvers, H., Simonsen, J. G., & Zantema, H. (2011a). Levels of undecidability in rewriting. *Information and Computation, 209*(2), 227–245.

Endrullis, J., Geuvers, H., & Zantema, H. (2009b). Degrees of undecidability in term rewriting. In *Proceedings of International Workshop on Computer Science Logic (CSL 2009)*, Vol. 5771 of *LNCS* (pp. 255–270). Berlin: Springer.

Endrullis, J., Grabmayer, C., & Hendriks, D. (2008). Data-oblivious stream productivity. In Proceedings Conference (Ed.), *on Logic for Programming Artificial Intelligence and Reasoning (LPAR 2008), number 5330 in LNCS* (pp. 79–96). Berlin: Springer.

Endrullis, J., Grabmayer, C., & Hendriks, D. (2009c). Complexity of Fractran and productivity. In *Proceedings Conference on Automated Deduction (CADE 22)*, Vol. 5663 of *LNCS* (pp. 371–387).

Endrullis, J., Grabmayer, C., Hendriks, D., Klop, J. W., & de Vrijer, R. C. (2009d). Proving infinitary normalization. In *Postproceedings International Workshop on Types for Proofs and Programs (TYPES 2008)*, Vol. 5497 of *Lecture Notes in Computer Science* (pp. 64–82). Berlin: Springer.

Endrullis, J., Grabmayer, C., Hendriks, D., Klop, J. W., & van Oostrom, V. (2010c). Unique normal forms in infinitary weakly orthogonal rewriting. In *Proceedings of 21st International Conference on Rewriting Techniques and Applications (RTA 2010)*, Vol. 6 of *Leibniz International Proceedings in Informatics* (pp. 85–102). Wadern: Schloss Dagstuhl.

Endrullis, J., Grabmayer, C., Hendriks, D., Isihara, A., & Klop, J. W. (2010b). Productivity of stream definitions. *Theoretical Computer Science, 411*, 765–782.

Endrullis, J., Hansen, H. H., Hendriks, D., Polonsky, A., & Silva, A. (2013a). A coinductive treatment of infinitary rewriting. *CoRR*, abs/1306.6224.

Endrullis, J., & Hendriks, D. (2011). Lazy productivity via termination. *Theoretical Computer Science, 412*(28), 3203–3225.

Endrullis, J., Hendriks, D., & Klop, J. W. (2010d). Modular construction of fixed point combinators and clocked Böhm trees. In *Proceedings of Symposium on Logic in Computer Science (LICS 2010)* (pp. 111–119). Edinburgh: IEEE Computer Society.

Endrullis, J., Hendriks, D., & Klop, J. W. (2012a). Highlights in infinitary rewriting and lambda calculus. *Theoretical Computer Science*, *464*, 48–71.

Endrullis, J., Hendriks, D., Klop, J. W., & Polonsky, A. (2012b). Discriminating lambda-terms using clocked Böhm trees. In *Logical methods in computer science* (In print).

Endrullis, J., Hendriks, D., Klop, J. W., & Polonsky, A. (2013b). *Clocked lambda calculus*. In Mathematical structures in computer science: Accepted for publication.

Endrullis, J., & Polonsky, A. (2013). Infinitary rewriting coinductively. In *Proceedings of Types for Proofs and Programs (TYPES 2012)*, Vol. 19 of *Leibniz International Proceedings in Informatics* (pp. 16–27). Wadern: Schloss Dagstuhl.

Geser, A., & Zantema, H. (1999). *Non-looping string rewriting. ITA*, *33*(3), 279–302.

Hindley, J. R., & Seldin, P. (2008). *Lambda-calculus and combinators*. Cambridge: Cambridge University Press.

Jacopini, G. (1975). A condition for identifying two elements of whatever model of combinatory logic. In *λ-Calculus and Computer Science Theory*, Vol. 37 of *Lecture notes in computer science* (pp. 213–219). Berlin: Springer.

Kennaway, J. R., & de Vries, F.-J. (2003). *Infinitary rewriting, Chap. 12*. Cambridge: Cambridge University Press. in Terese (2003).

Kennaway, J. R., Klop, J. W., Sleep, M. R., & de Vries, F.-J. (1995). Transfinite reductions in orthogonal term rewriting systems. *Information and Computation*, *119*(1), 18–38.

Kennaway, J. R., Klop, J. W., Sleep, M. R., & de Vries, F.-J. (1997). Infinitary lambda calculus. *Theoretical Computer Science*, *175*(1), 93–125.

Kennaway, J. R., van Oostrom, V., & de Vries, F.-J. (1999). Meaningless terms in rewriting. *The Journal of Functional and Logic Programming, 1*.

Ketema, J., Klop, J. W., & van Oostrom, V. (2005). Vicious circles in orthogonal term rewriting systems. In S. Antoy & Y. Toyama (Eds.), *Proceedings of 4th International Workshop on Reduction Strategies in Rewriting and Programming (WRS'04)*, Vol. 124(2) of *Electronic Notes in Theoretical Computer Science* (pp. 65–77). Elsevier Science.

Klop, J. W. (1980). *Combinatory reduction systems*, Vol. 127 of *Mathematical centre tracts*. Amsterdam: Mathematisch Centrum.

Klop, J. W. (1992). Term rewriting systems. In *Handbook of logic in computer science* (Vol. II, pp. 1–116). Oxford: Oxford University Press.

Klop, J. W., & de Vrijer, R. C. (2005). Infinitary normalization. *We will show them: Essays in Honour of Dov Gabbay (2)* (pp. 169–192). London: College Publications.

Middeldorp, A. (1997). Call by need computations to root-stable form. In *Proceedings of Symposium on Principles of Programming Languages (POPL 1997)* (pp. 94–105). New York: ACM.

Schönfinkel, M. (1924). Über die Bausteine der mathematischen Logik. *Mathematische Annalen*, *92*, 305–316.

Scott, D. S. (1975). Open problem. In C. Böhm (Ed.), *Lambda calculus and computer science theory*, Vol. 37 of *LNCS* (p. 368).

Smullyan, R. M. (1985). *To Mock a Mockingbird, and other logic puzzles*. New York: Alfred A. Knopf.

Smullyan, R. M. (1994). *Diagonalization and self-reference*. Oxford: Clarendon Press.

Terese. (2003). *Term rewriting systems*, Vol. 55 of *Cambridge tracts in theoretical computer science*. Cambridge: Cambridge University Press.

van der Poel W. L., approximately 1975. Personal communication.

van der Poel, W. L. (1992). *The mechanization of lambda calculus*. Reports of the Faculty of Technical Mathematics and Informatics, Delft University of Technology.

van Heijenoort, J. (1967). *From Frege to Gödel: A source book in mathematical logic, 1879–1931*. Harvard University Press: Source books in the history of the sciences. Cambridge.

Visser, A. (1981). *Aspects of diagonalization and probability*. Ph.D. thesis, Utrecht University.

Waldmann, J. (1997). Nimm Zwei. Technical Report IR-432, Vrije Universiteit Amsterdam.

Waldmann, J. (1998). *The combinator S*. Ph.D. thesis, Friedrich-Schiller-Universität Jena.

Waldmann, J. (2000). The combinator S. *Information and Computation*, *159*(1–2), 2–21.

Zachos, E. (1978). *Kombinatorische Logik und S-Terme*. Ph.D. thesis, Berichte des Instituts für Informatik, Eidgenössische Technische Hochschule Zürich.

Zachos, S., Cheilaris, P., & Ramirez, J. (2011). Checking in linear time if an S-term normalizes. In *8th Panhellenic Logic Symposium*.

Zantema, H. (1995). Termination of term rewriting by semantic labelling. *Fundamenta Informaticae*, *24*, 89–105.

Chapter 6
Some Tweets About Mockingbirds

Rick Statman

For Ray Smullyan:

Hush, little baby, don't say a word,
Raymond's gonna buy you a mockingbird.

If that mockingbird won't mock,
Raymond's gonna buy you some Google stock!

Abstract In his fascinating book "To Mock a Mockingbird", Ray Smullyan asks whether a fixed point combinator can be constructed from the Bluebird B, the Identity Bird I, and the Mockingbird M (sometimes called "little omega"). This is still not known.The mockingbird is only a special kind of mimid. Mimids are those birds which upon hearing the name of x, say what x would say upon hearing the name of some other y. In this note we shall characterize the set of mimids that occur in the Bluebird/Mockingbird flock.

Keywords Mockingbird · Combinator · Mimid · Isochrony

6.1 Introduction

In his fascinating book "To Mock a Mockingbird", Ray Smullyan asks whether a fixed point combinator can be constructed from the Bluebird B, the Identity Bird I, and the Mockingbird M (sometimes called "little omega"). This is still not known. In this note we shall consider some related questions.

The B, I monoid was first singled out as a fragment of lambda calculus by Church (1941) in connection with the word problem. It comes up naturally as a subset of Church's basis $B, I, C*, W*, K$, where $M = W* = WI$, and was further studied by Curry (1958) in his work on the inter-definability of combinators; in particular,

R. Statman (✉)
Department of Mathematical Sciences, Carnegie Mellon University,
Pittsburgh, PA 15213, USA
e-mail: statman@cs.cmu.edu

M. Fitting and B. Rayman (eds.), *Raymond Smullyan on Self Reference*,
Outstanding Contributions to Logic 14, https://doi.org/10.1007/978-3-319-68732-2_6

with reference to compositive, permutative, duplicative, and selective effect. This was taken up years later by the author (Smullyan 1986).

A fixed point combinator F is a combinator satisfying

$$Fx = x(Fx).$$

Now B and M are sufficient to define fixed points

$$M(BxM) = BxM(BxM) = x(M(BxM))$$

and in the presence of W, we have a fixed point combinator

$$B(BMW)B.$$

But, without "permutative effect", it is unclear whether a fixed point combinator can be defined from B, I and M alone. Note that $BM(B(BM)B)$ has the right "Bohm tree" to be a fixed point combinator but it is not one. This can be proved directly.

The question was studied by Wos and McCune (1996), and in Statman (1993) we proved that there is no such F satisfying

$$Fx \twoheadrightarrow x(Fx)$$

where $\twoheadrightarrow$ is ordinary (weak beta) reduction.

6.2 Mimids

The mockingbird is only a special kind of mimid. Mimids are those applicative combinations, X, of B, I, and M, such that there exists an applicative combination of B, I, M, and x, X, with

$$Fx = xX.$$

In other words, F, upon hearing the name of x, says what x would say upon hearing the name of some other bird X. In this note we shall characterize the set of mimids in an interesting way.

We shall assume for the most part that the reader is familiar with classical lambda calculus and combinatory logic as in (Church 1941). B, I and M can be regarded as constants with (weak beta) reduction

(I) $Ix \rightarrowtail x$
(B) $Bxyz \rightarrowtail x(yz)$,
(M) $Mx \rightarrowtail xx$

and the corresponding conversion relation, $\leftrightarrow$, or as lambda terms

$I = \lambda x.\, x,$

$B = \lambda xyz.\, x(yz),$ and

$M = \lambda x.\, xx$

with beta, or beta-eta reduction. Although Smullyan's question is about constants with weak-beta, it will be helpful to consider lambda terms with beta-eta. We will need to know a few things about $\leftrightarrow$.

Fact 1

(i) $\rightarrowtail$ satisfies the standardization theorem.
(ii) $\leftrightarrow$ satisfies the Church-Rosser theorem.

Lemma 6.2.1 *Suppose that X is a combination of B, I and M and $Xx_1 \dots x_k \leftrightarrow x_i X_1 \dots X_l$, then $i = 1$ and X has $\rightarrowtail$ normal form.*

Remark: This lemma is slightly stronger than the one stated in Statman 1993.

Proof Suppose that $X\, x_1 \dots x_k \leftrightarrow x_i X_1 \dots X_l$. By the Church-Rosser and standardization theorems

$$X x_1 \dots x_k \twoheadrightarrow x_i X_1 \dots X_l$$

by a head reduction (possibly for different X_j) The proof is by induction on the length of a head reduction. We first consider the case of a head reduction of X. This case follows immediately from the induction hypothesis. Next we consider the case that X is in head normal form

Case 1: $X = B, I, M$. This case is trivial.
Case 2: $X = BUV$. We have

$$BUVx_1 \dots x_k \rightarrowtail U(Vx_1)x_2 \dots x_k.$$

Now we can simulate the reduction

$$U(Vx_1)x_2 \dots x_k \twoheadrightarrow x_i X_1 \dots X_l$$

by $Ux_1x_2 \dots x_k$ until Vx_1 comes to the head or the reduction ends. But by the induction hypothesis the second option is impossible. Thus by induction hypothesis U has a normal form. Now when Vx_1 comes to the head we have

$$Vx_1V_1 \dots V_m \twoheadrightarrow x_i X_1 \dots X_k.$$

Now we can simulate this head reduction with

$$Vx_1y_1 \dots y_m$$

until one of the V_j comes to the head or the reduction ends. By the induction hypothesis the first alternative is impossible so by the induction hypothesis V has a normal form.

Case 3: $X = BU$. We have

$$BUx_1 \dots x_k \rightarrowtail U(x_1x_2)x_3 \dots x_k.$$

This case is similar to case 2. □

It is worth noting that this lemma remains true if M is replaced by W.

We need some notation $X^n x =: X(\dots(Xx)\dots) n$ occurrences of X and a definiton.

Definition of monomials and polynomials;
$B^n B$ and $B^n M$ are monomials.
Each monomial is a polynomial.
If X and Y are polynomials then BXY is a polynomial.

We have not defined I to be a polynomial. Equivalently, I could be regarded as the trivial polynomial.

Certain $\rightarrowtail$ normal terms like $B(Bxy)$ are not polynomials, but they are beta-eta convertible to polynomials.

Fact 2 $(B, I$monoid);

(i) $Bx(Byz)$ beta-eta conv. $B(Bxy)z$
(ii) BI beta-eta conv. I
$Bx\ I$ beta − eta conv. x
(iii) $B(Bxy)$ beta-eta conv. $B(Bx)(By)$
(iv) $BB(Bx)$ beta-eta conv. $B(B(Bx))B$

Proof Only (iii) and (iv) may be unfamiliar. For (iii), just apply (i) to $B(Bxy)z$ to get $Bx(Byz) \leftarrowtail B(Bx)(By)z$. For (iv), $BB(Bx)y \rightarrowtail B(Bxy)$ which by (iii) beta-eta conv. $B(Bx)(By) \leftarrowtail B(B(Bx))By$. □

Corollary 6.2.1 *If X is $\rightarrowtail$ normal then X beta-eta conv. to a polynomial or I.*

Indeed, if F is normal and a mimid then the result of repeated application of (ii) and (I) to F is also normal, I free, and a mimid. The converse is annoyingly false since F may be a mimid but $B(BI)F$ is not. So, we adjoin the reductions

(I') $BI \rightarrowtail I$
$(I")$ $BxI \rightarrowtail x$

to weak beta reduction together with the remark that the Church-Rosser theorem, but not standardization, is preserved.

Since B is associative by (i), we shall write BXY as $X * Y$ using the customary infix "multiplicitive" notation.

If F is a mimid then, by Lemma 1, F has a $\rightarrowtail$ normal form and so, by Fact 2, F beta-eta converts to a polynomial. Conversely,

Lemma 6.2.2 *If F has a $\rightarrowtail$ normal form and is I free and Fx beta-eta converts to xX then F is a mimid.*

Proof If Fx beta eta converts to xX then by the Church-Rosser theorem and eta postponement there exists an eta expansion Y of xX such that Fx beta reduces to Y. By the standarization theorem there exists a standard reduction to Y. A standard reduction can be thought of as a tree of head reductions. The proof is by induction on the depth of this tree. We may assume that

$$Y = \lambda y_1 \dots y_t . xXY_1 \dots Y_t$$

(not the same X, but one that eta reduces to X) where y_i is not in X and Y_i eta reduces to y_i. Now we can simulate the head reduction part of the standard reduction $Fxy_1 \dots y_t$ to $xXY_1 \dots Y_t$ by $\rightarrowtail$. If no y_i is ever an argument of a head redex then we are done. Otherwise consider the first such redex. The possible forms are

$BUy_1y_2 \dots y_t$, and
$BUVy_1y_2 \dots y_t$.

The first of these prevents eta reduction of y_1. The last yields $U(Vy_1)y_2 \dots y_t$. Since V contains x, U cannot contain x, and Vy_1 must come to the head of the term in the head reduction without copying y_1. So the normal form of U is an applicative combination of B's, indeed the normal form of U must be B^kB for some k. But this is impossible at the last step. □

We shall assume that F is the polynomial

$$F_0 * \dots * F_m$$

using association of $*$ to the right $x * y * z = x * (y * z)$. Each monomial F_i has the form $B^{h(i)}G_i$, where $G_i = B$ or M. G_i is called the root of the monomial. We shall now exactly characterize which polynomials are mimids.

Fact 3 If the polynomial F is a mimid then F_0 is M.

Proof Let $G = F_1 * \dots * F_m$. Then $Fx \rightarrowtail F_0(Gx)$ and so by the Church-Rosser and standardization theorems we can assume that there is a head reduction from $F_0(Gx)$ to xX. We say "can assume" here because the X may change to an equivalent ($\leftrightarrow$) one. Thus $h(0) = 0$ and the root is M. □

Now for the polynomial F we define a sequence $f = f_1 \dots f_m$ of parentheses $f_i = L$ (left) or $f_i = R$ (right) according to whether the root of F_i is M or B resp. Let W be the set of i s.t. $f_i = L$ and $E = \{1, \dots, m\} - W$. Recall that a proper pairing of parentheses is a bijection $p : W \to E$ such that $p(i) > i$ and for $i, j : W$ we have $i < j$ implies $p(j) < p(i)$.

Lemma 6.2.3 *If F is a mimid then f has a proper pairing.*

Proof If X is an applicative combination of B, M, and x we define $\#X$ as follows. X can be thought of as a binary tree with B, M and x at its leaves. Each vertex, except the root, has a unique incoming edge, and each vertex, except a leaf, has a unique outgoing left edge and a unique outgoing right edge. Each leaf l has a unique path from the root with $\#l$ outgoing left edges. $\#X$ is # of the leftmost occurrence of x. So $\#(Fx) = 0$ and $\#(Gx)(Gx) = 1$. Now suppose that F is a mimid and we have a head reduction $Fx \twoheadrightarrow xX$. Assume that one of the terms J in this reduction has the form $G_i x H_1 \ldots H_n$ where $G_i = F_i * \ldots * F_m$; so, $G_1 = G$. We shall split the reduction into segments. Observe that if the root of F_i is B then F_i beta-eta reduces to

$$\lambda x_0 \ldots x_{h(i)} yz.\, x_0 \ldots x_{h(i)}(yz)$$

and if the root of F_i is M then F_i beta-eta reduces to

$$\lambda x_0 \ldots x_{h(i)}.\, (x_0 \ldots x_{h(i)})(x_0 \ldots x_{h(i)}).$$

So in the first case we have $\#J = n > h(i) + 1$ and the segment we consider consists of

$$J \ldots G_{i+1} x H_1 \ldots H_{h(i)}(H_{h(i)+1} H_{h(i)+2} \ldots H_n)$$

where $\#(G_{i+1} x H_1 \ldots H_{h(i)}(H_{h(i)+1} H_{h(i)+2} \ldots H_n) = n - 1$.

In the second case we have $\#J = n > h(i) - 1$ and the segment consists of

$$J \ldots G_{i+1} x H_1 \ldots H_{h(i)}(G_{i+1} x H_1 \ldots H_{h(i)}) H_{h(i)+1} \ldots H_n$$

where $\#(G_{i+1} x H_1 \ldots H_{h(i)}(G_{i+1} x H_1 \ldots H_{h(i)}) H_{h(i)+1} \ldots H_n) = n + 1$. Now for the sequence $g = g_1 \ldots g_m$ where

$$g_i = \#(G_i x H_1 \ldots H_n) = n$$

we have
$g_1 = 1$
$g_{m+1} = 1$
$g_i > 0$ for $i = 1, \ldots, m+1$
$g_{i+1} = g_i + 1$ if $f_i = L$
$g_{i+1} = g_i - 1$ if $f_i = R$.

Now it is part of the folklore of elementary combinatorics that the sequence $f_1 \ldots f_m$ admits a proper pairing if and only if the sequence g satisfies these conditions. □

For the sequence $h = h(1) \ldots h(m)$ we say g majorizes h if $f_i = L$ implies $h(i) - 1 < g_i$ and $f_i = R$ implies $h(i) + 1 < g_i$. In the proof of Lemma 2 we have shown that

Corollary 6.2.2 *If F is mimid then g majorizes h.*

Example 1: h is identically 0. Every proper pairing of $2n$ parens can be obtained from the pairing $LRLR \dots LR$ by repeating the following transformation; given $\dots L \dots L \dots R \dots R \dots$ where the shown parens are paired and no paren strictly in $L \dots L$ is paired to one in $R \dots R$, replace the pairing by $\dots L \dots R \dots L \dots R \dots$ This can be seen by observing that the sequence g_i drops in the product ordering.

This was probably known to Euler but I can find no reference. This enumerates all the cases for Example 1.

Example 1 (cont'd): h is identically 0. Define X_m recursively so that Fx beta-eta conv. $x\ X_m$. $X_0 := x$. For $m > 0$, since f has a proper pairing we can write $G_1 = H * M * B * \dots * B$ for $k > 0$ occurrences of B. So, $G_1 x$

beta-eta conv. $H(M(B^k x))$
beta-eta conv. $H((B^k x)(B^k x))$
beta-eta conv. $H(B^{k-1}(x * Bx))$
beta-eta conv. $H * B^{k-1}(x * Bx)$.

Now $H * B^{k-1}$ has a proper pairing for $m = m - 2$ so we have $M * H * B^{k-1}x$ beta-eta conv. xX_{m-2}. Substituting $x * Bx$ for x gives $X_m := x * [x * Bx/x]X_{m-2}$. Now we can describe X_m (up to association of $*$) as follows. X_m is a product of $2^{m/2}$ terms, the ith term of which, for $i = 0, \dots, 2^{m/2} - 1$, is

$$B^{l(i)}x$$

where $l(i)$ is the l_1 norm of i (Hamming distance).

Example 2: h is maximal i.e. $f_i = L$ implies $h(i) = g_i$ and $f_i = R$ implies $h(i) + 2 = g_i$. In this case no $F_i = M$ except for $i = 0$ since if $F_i = M$ then $g(i) > 0$ and $g(i + 1) = g(i) + 1$ which is the same for M replaced by BM. Indeed if $F_i * F_{i+1} = B * BH$ then we can replace this by the beta-eta equivalent $B(BH) * B$. This will not be maximal but using the inequivalent $B(BH) * BB$ the result is maximal. After a series of such transformations we obtain a maximal $F' = M * BH * B * \dots * B$ where $H * \dots$ is maximal.

Lemma 6.2.4 *If $F_0 = M$, f admits a proper pairing, and g majorizes h then F is a mimid.* □

Proof Now suppose that we have a head reduction $Fx \rightarrowtail \dots$. Assume that one of the terms J in this reduction has the form $G_i x H_1 \dots H_{g(i)}$ where $G_i = F_i * \dots * F_m$. If the root of F_i is B then $h(i) + 1 < g(i)$ so we have $J \rightarrowtail G_{i+1} x H_1 \dots (H_{h(i)+1} H_{h(i)+2}) \dots H_{g(i)}$ where $g(i + 1) = g(i) - 1$. If the root of F_i is M then $J \rightarrowtail G_{i+1} x H_1 \dots H_{h(i)} (G_{i+1} x H_1 \dots H_{h(i)}) H_{h(i)+1} \dots H_{g(i)}$ where $g(i + 1) = g(i) + 1$. In particular, when $i = m$, since $g(m + 1) = 1$, we have $g(m) = 2$, $G_m = B$, and the next term in the reduction is $x(H_1 H_2)$. Thus F is a mimid.

We can now characterize the mimids.

Theorem 6.2.1 *F is a mimid if and only if F has a $\rightarrowtail$ normal form which beta-eta converts to an I free polynomial beginning with M such that the sequence f admits a proper pairing and the sequence g majorizes the sequence h.*

6.3 Isochrony

In order to sing the name of FG we must sing the name of G to the bird F, which we should have in hand. In order to have the bird F in hand we must know its name; that is, we must have its name sung by a bird we already have in hand. Of course we have B, M, and I at hand to begin with. In addition, many of these birds that need to be gathered can be gathered in parallel, provided that they can be named independently. Finally, some of these birds may, in fact, be the same bird, but recognizing this appears to be a very hard problem and we must skip it here. The time it takes before we can sing the name of the bird H, $\#H$, can be defined $\#B = \#M = \#I = 0$ and $\#(FG) = \max\{1 + \#F, \#G\}$.

A similar story can be told about space. In order to sing the name of FG we must have a perch for F and all the perches we need to sing the name of G. Thus, the number of perches needed to sing the name of the bird H. $\$H$ can be defined by $\$B = \$M = \$I = 0$ and $\$(FG) = \max\{\$F, 1 + \$G\}$. Below we shall consider #; we claim similar results hold for \$.

As in the proof of Lemma 3, applicative combinations F of B, M and I can be thought of as binary trees with the atoms B, M, and I at the leaves. Each vertex, except the root, has a unique incoming edge, and each vertex, except a leaf, has a unique outgoing left edge and a unique outgoing right edge. Each leaf l has a unique path from the root with $\#l$ outgoing left edges. Thus

$$\#H = \max\{\#l \mid l \text{ is a leaf of } H\}.$$

Indeed, we shall show that for any n there are birds F which require $\#H > n$ for any $H = F$. Somewhat surprisingly, for general combinatory logic (under any basis) this is not the case.

Theorem 6.3.1 *For combinatory logic under any basis there exists n such that for any F there exists $H = F$ with $\#H < n$.*

Proof By Kleene's enumeration theorem for the combinator case there exists a combinator E, an combinatory Church numeral 0 for zero, and a combinatory successor N for the arithmetic successor function such that if m is the Godel number of the term F then

$$E(N^m 0) \twoheadrightarrow F$$

and we have $\#(E(N^m 0)) = \max\{1 + \#E, 1 + \#N\}$ which does not depend on m. □

Let $M_0 = M$ and $M_{m+1} = M_m M$. We shall show that for any $F = M_{m+1}$ we have $\#F > m$. Toward this end we will want to consider more reductions (see (iii) above)

(B') $B(Bxy) \rightarrowtail B(Bx)(By)$
and less #

$$\#H = \max\{\#l \mid l \text{ is a leaf of } H \text{ labeled } M\}.$$

Here we note that only (M) reductions can increase the new #. We now have six reduction rules so we will have to be explicit at every turn about which ones we are using.

Fact 4 In $(B) + (M) + (I)$, if $F = M_m$ then $F \twoheadrightarrow M_m$. Moreover, F can be reduced to M_m by an alternating sequence of (M) reductions and (B), (I) reductions to $(B) + (I)$ normal form. We shall measure a reduction R of F to M_m by the number r of (M) reductions. We shall now observe the analogue of the "parallel moves lemma" for this context.

Lemma 6.3.1 *Suppose that $F \twoheadrightarrow M_m$ in $(B) + (M) + (I)$ by R then*

(i) If $F \rightarrowtail F'$ in $(B') + (I') + (I'')$ then $F' \twoheadrightarrow M_m$ in $(B) + (M) + (I)$ by a reduction R' with $r' < r + 1$
(ii) If $F \rightarrowtail F'$ by the contraction of an innermost (M) redex then $F' \twoheadrightarrow M_m$ in $(B) + (M) + (I)$ by a reduction R' with $r' < r + 1$.

Proof The proof uses the notion of the residuals of a redex ([1]). In the case of combinators these are disjoint. Each "step" in the reduction of F corresponds, by a complete reduction of all residuals of the redex contracted to obtain F', to at most one "step" in the reduction of F'. Here, of course, multiple consecutive $(B) + (I)$ steps to $(B) + (I)$ normal form count as one. □

Lemma 6.3.2 *If $M F_1 \dots F_n \twoheadrightarrow M_m$ then F_1 has a $(B) + (M) + (I)$ normal form.*

Proof We have $F_1 F_1 \dots F_n \twoheadrightarrow M_m$ by a standard reduction. If the second occurrence of F_1 comes to the head then F_1 has a normal form by Lemma 1. Otherwise, the second occurrence of F_1 has a trace which is a subterm of a term which $\twoheadrightarrow M$ and thus F_1 is again normalizable. □

Theorem 6.3.2 *If $F = M_{m+1}$ in $(B) + (M) + (I)$ then $\#F > m$.*

Proof Let F have smallest $\#F$ such that $F = M_{m+1}$ and a $(B) + (M) + (I)$ reduction R with smallest r for this $\#F$. Then we can assume F is $(B) + (I)$ and has the form

$$M F_1 \dots F_n.$$

By Lemma 4 we can assume that F_1 is a polynomial associated to the right since $B(Bxy)z \rightarrowtail B(Bx)(By)z \twoheadrightarrow Bx(Byz)$. We distinguish two cases.

Case 1: F_1 does not contain M. This case contradicts either the choice of $\#F$ or the choice of r.
Case 2: F_1 contains M. Now F_1 is a polynomial so $\#F_1 = 1$. In order for the choices to be not contradicted we must have

$$\#F_1F_1\ldots F_n = 1 + \#F.$$

Now we can write $F_1 = G * B^kM * H$ where G is a monomial with no occurrence of M. Thus for $l < n$

$$F_1F_1\ldots F_n \rightarrowtail\!\!\!\!\rightarrow (B^kM * H)F_1G_1...G_l$$

$$\rightarrowtail\!\!\!\!\rightarrow M(HF_1G_1\ldots G_p)G_{p+1}\ldots G_l$$

by only (B) and (I) reductions. By Lemma 6 $HF_1G_1\ldots G_p$ has a $(B)+(M)+(I)$ normal form. This contradicts the choice of r. □

References

Barendregt, H. (1983). *The lambda calculus*. Amsterdam: North Holland.

Church, A. (1941). *The calculi of lambda conversion*. Princeton: Princeton University Press.

Curry, H. B. (1958). *Combinatory logic* (Vol. 1). Amsterdam: North Holland.

Smullyan, R. (1985). *To mock a Mockingbird*. New York: Knopf.

Statman, R. (1986). On translating lambda terms into combinators. In *Proceedings of the Logic in Computer Science Conference I.E.E.E.*

Statman, R. (1991). Freyd's hierarchy of combinator monoids. In *Proceedings of the Logic in Computer Science Conference I.E.E.E.*

Statman, R. (1993). Some examples of non-existant combinators. *Theoretical Computer Science*, *121*, 441–448.

Statman, R. (2014). Near semi-rings and lambda calculus, *12th International Conference on Typed Lambda Calculus and Applications*, July 14–17, 2014. Vienna Austria (to appear in Lecture Notes in Computer Science).

Wos, L., & McCune, W. (1996). The absence and presence of fixed point combinators. *Automated Deduction - CADE-13 Lecture Notes in Computer Science*. p. 1194.

Chapter 7
Knights, Knaves, Truth, Truthfulness, Grounding, Tethering, Aboutness, and Paradox

Stephen Yablo

Abstract Knights always tell the truth; Knaves always lie. Knaves for familiar reasons cannot coherently describe themselves as liars. That would be like Epimenides the Cretan accusing all Cretans of lying. Knights do not *intuitively* run into the same problem. What could prevent a Knight from truly reporting that s/he always tells the truth? Standard theories of truth DO prevent this, however, for such a report is self-referentially ungrounded. Standard theories have a problem, then! We try to fix it.

7.1 Knights and Knaves

Knights, as we know, always tell the truth; knaves always lie. Knight and knave puzzles ask us to figure out who is who on the basis of their answers to cleverly contrived questions. For instance,

> *A*, *B*, and *C* were standing together in a garden. A stranger passed by and asked *A*, "Are you a knight or a knave?" *A* answered, but rather indistinctly, so the stranger could not make out what he said. The stranger then asked *B*, "What did *A* say?" *B* replied, "*A* said that he is a knave." At this point the third man, *C*, said, "Don't believe *B*; he is lying!" The question is, what are *B* and *C*? (Smullyan 1986, 20)

Smullyan begins by observing that

> It is impossible for either a knight or a knave to say, "I'm a knave," because a knight wouldn't make the false statement that he is a knave, and a knave wouldn't make the true statement that he is a knave.

He concludes on this basis that *B*, since he is lying about what *A* said, is a knave; *C* must be a knight since he is right about *B*; *A*'s status cannot be determined.

A variant of the puzzle can be imagined in which *B* replies, not "*A* said he was a knave," but "*A* said that he was a knight." *B* speaks the truth, for knights and knaves *both* say, "I am a knight"—knights because "I am a knight" is true in their mouths,

S. Yablo (✉)
Massachusetts Institute of Technology, Cambridge, MA, USA
e-mail: yablo@mit.edu

M. Fitting and B. Rayman (eds.), *Raymond Smullyan on Self Reference*,
Outstanding Contributions to Logic 14, https://doi.org/10.1007/978-3-319-68732-2_7

and knaves because it is false in theirs. Since his description of *A* is true, *B* must be a knight. *B* might equivalently have replied that *A* said he always told the truth, for that is the kind of speech behavior that is definitive of a knight.

Straightforward as this reasoning appears, there is, to go by current theories of truth and self-reference, something badly wrong with it. Knights cannot, on current theories, truly describe themselves as always telling the truth. That the problem is not apparent even to veteran paradox-mongers (see below) is a datum in need of explanation. This paper seeks mainly to *explain* the problem. But we will take a shot, toward the end, at addressing it.

7.2 Russell and Moore

The Smullyan puzzle recalls a remark of Kripke's about Russell's sense of, or radar for, paradox. Russell asked Moore, *Do you always tell the truth?* Moore replied that he didn't. Russell

> regarded Moore's negative reply as the sole falsehood Moore had ever produced. Surely no one had a keener nose for paradox than Russell. Yet he apparently failed to realize that if, as he thought, all Moore's other utterances were true, Moore's negative reply was not simply false but paradoxical (Kripke 1975, 691–692)

Why paradoxical? Assume first that the statement is false. Then Moore does sometimes lie, in which case the statement is true after all. If on the other hand it is true, then Moore never lies, in which case the answer he gives Russell is just incorrect. A statement that cannot consistently be assigned either truth-value is normally considered paradoxical. "Even the subtlest experts," Kripke says, "may not be able to avoid utterances leading to paradox."

7.3 Moore Be(k)nighted

And yet, there seems to be something right about Russell's claim that Moore spoke falsely. How else are we to describe the situation, if we cannot call Moore's mea culpa a lie? All of Moore's other statements are true, we're supposing. His statement *I sometimes lie* has, therefore, no basis in fact. To call it untrue seems like our only option if we want to give voice to this observation. And yet to call it untrue is self-refuting.

Russell may have put his point in an unnecessarily paradoxical way. Perhaps he meant, not that Moore's *actual* statement, *I sometimes lie*, was untrue, but that the opposite statement, *I always tell the truth*, would have been true, had he made it. That *I (Moore) always speak the truth* would have been true does seem intuitively rather similar to what Russell alleges, viz. that *I (Moore) sometimes lie* is false. One feels that had Moore said instead that he never lied, or that all his statements were true, he would have spoken truly. An honest person ought to be able to assert their own honesty![1] And that is what Moore would be doing in the imagined scenario.

Where does this leave us? Even if Moore did not *lie*, when he said *I sometimes lie*, Russell can be forgiven, so it seems, for thinking that he did. The judgment is forgivable for it is easily confused with (what seems so far to be) the *correct* judgment that Moore would have done better to say, *I always tell the truth*, since he would then have been speaking truly. This seems like a very satisfactory resolution. It allows us to agree with Kripke that Russell misconstrued a paradox as a lie, while also agreeing with Russell that Moore's reply to *Do you ever lie?* was an unforced error, in this sense: the answer he did give (*YES*) was indefensible, while the answer he didn't give (*NO*) would have been true. Russell had the right idea, on this interpretation; he simply didn't say it right.

7.4 The Problem

To explain the false-seemingness of *I sometimes lie* as reflecting the truth of *I never lie* seems like a satisfactory resolution. But the plot now begins to thicken. Granted that *I (Moore) never lie* is not paradoxical, there is still the problem of seeing why it should be regarded as *true*. It is after all self-referential; it attributes truth to itself. Statements like that may not be consigned to the *first* circle of hell, but they *are* often sent to the second.

There's an intuitive aspect to this and a technical aspect. The intuitive aspect is as follows. You all know of the Liar sentence L, which describes itself as untrue ($L = \neg T(L)$). The Liar cannot consistently be regarded either as true or as false; that is more or less what it means to be paradoxical. Paradox is not the only form of semantic pathology, however, as remarked by Kripke:

> It has long been recognized that some of the intuitive trouble with Liar sentences is shared with such sentences as
>
> (K) K is true
>
> which, though not paradoxical, yield no determinate truth conditions (Kripke 1975, 693)

Where the Liar can consistently be assigned *neither* truth-value, the Truth-Teller K can consistently be assigned *either*. Suppose we call it true; then what it says is the case; and so it deserves the description we gave it. Likewise if we call it false. We can assign it whatever truth-value we like and that assignment will bear itself out. Borrowing a term from Kripke, the Truth-Teller is not *paradoxical* (overdetermined) but *indeterminate* (underdetermined).

Return now to *Everything I say is true*. I will call it the Truthfulness-Teller, because the speaker (Moore, we suppose) is declaring himself to be generally truthful, and

[1] Self-identified knights are the group Smullyan admires the most. If they were talking nonsense, he would have noticed it.

write it *H*, for honesty. *H* is, it may seem, in the same boat as the Truth-Teller *K*, assuming that the speaker's other statements are true. It is equivalent after all to *Everything else I say is true, and this statement too is true*. If we postulate that Moore lies when he calls *I always tell the truth* false, the postulate is self-supporting. What the sentence says really is false, on the assumption of its falsity, because it describes itself as true. If we assume for argument's sake that it is true, that assessment is self-supporting too.

So, the Truthfulness-Teller is true on the assumption of its truth, and false on the assumption of its falsity. A sentence that can consistently be supposed either true or false, compatibly with the non-semantic facts, is, it seems, indeterminate. The Truthfulness-Teller was *introduced*, though, precisely as a *truth* that Moore had available to him to utter, when he said instead that he was not always truthful, thus involving himself in paradox. The statement's truth was indeed proposed as what lent the appearance of falsity to *I sometimes lie*.

That's the intuitive aspect. The technical aspect is that if you look at the various formal truth theories that have been proposed — Tarski's, Kripke's, the Herzberger/Gupta theory, McGee's theory, Field's theory—not a single one of them supports the thought that Moore could truthfully have declared himself to be honest. Kripke's theory doesn't, for instance, because a sentence attributing truth to itself is *ungrounded* in the manner of the Truth-Teller and the Liar. Gupta's theory doesn't make *I never lie* true, for it is stably true in some revision-sequences but not others. Herzberger's version of the revision theory makes the Truthfulness-Teller just *false*, for it assigns the truth-predicate, initially, an empty extension, a setback from which *I never lie* cannot recover.[2]

7.5 Kripke and Dependence Trees

There are really two puzzles here. One, the comparative puzzle, asks why the Truthfulness-Teller should seem truer than the Truth-Teller, despite making a stronger claim. The absolute puzzle asks why the Truthfulness-Teller should be true full stop. Insofar as the first puzzle is to do with *H* seeming less grounded than *K*, and the second with *H* being ungrounded full stop, the natural context for either is Kripke's theory, for it was Kripke who put grounding at the center of the things.[3]

To appreciate how the theory works, let's associate with each sentence *P* two "attributions" $|P|^{\mathbf{t}}$ and $|P|^{\mathbf{f}}$, one assigning truth to *P*, the other falsity. A relation Δ on the set of attributions is a *dependence* relation iff it satisfies these conditions:

---

[2]Kripke does allow ungrounded sentences to be *intrinsically* true: true in a fixed point none of whose assignments are reversed in other fixed points. But the Truthfulness-Teller cannot claim that lesser status either, for there are fixed points in which it is uniquely false.

[3]Kripke cites Herzberger (1970). See also Davis (1979), Hazen (1981), Yablo (1982), and Yablo (1993). For the relation to grounding in set theory, see Mirimanoff (1917), Yuting (1953), Boolos (1971), Barwise and Etchemendy (1989), McLarty (1993), and Yablo (2006).

(a) if A is atomic, $|A|^{\mathbf{t}}$ and $|A|^{\mathbf{f}}$ bear Δ to nothing (written $\varnothing$)
(n) $|\neg P|^{\mathbf{t}}$ bears Δ to $|P|^{\mathbf{f}}$; $|\neg P|^{\mathbf{f}}$ bears Δ to $|P|^{\mathbf{t}}$
(d) $|P \vee Q|^{\mathbf{t}}$ bears Δ either to $|P|^{\mathbf{t}}$ or $|Q|^{\mathbf{t}}$; $|P \vee Q|^{\mathbf{f}}$ bears Δ both to $|P|^{\mathbf{f}}$ and $|Q|^{\mathbf{f}}$
(u) $|\forall x\, \varphi(x)|^{\mathbf{t}}$ bears Δ to $|\varphi(n)|^{\mathbf{t}}$ for each name n; $|\forall x\, \varphi(x)|^{\mathbf{f}}$ bears Δ to $|\varphi(n)|^{\mathbf{f}}$ for some particular name n
(t) $|T(A)|^{\mathbf{t}}$ bears Δ to $|A|^{\mathbf{t}}$; $|T(A)|^{\mathbf{f}}$ bears Δ to $|A|^{\mathbf{f}}$

P is *grounded-true* iff there is a dependence relation Δ such that every Δ-path starting from $|P|^{\mathbf{t}}$ leads to a fact—an atomic attribution $|A|^{\mathbf{t}}$ ($|A|^{\mathbf{f}}$) such that A really is true (false) in reality, as represented by the underlying model. Equivalently, $|P|^{\mathbf{t}}$ sits atop a factual Δ tree—a dependence tree all of whose branches terminate in facts. The rules in tree form:

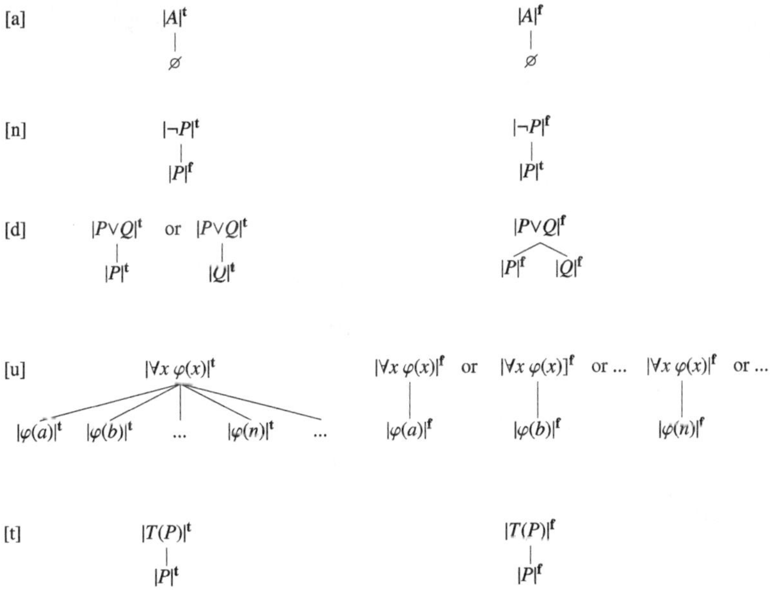

One way to define grounded-truth is in terms of trees whose branches terminate in *facts*: atomic attributions in which the sentence really does have the indicated truth-value. A different, but equivalent, way, uses *decorated* trees whose attributions are marked ✓ if they're factual and ✗ if they conflict with the facts. To get a decorated tree from a plain one, one starts by tagging terminal nodes with ✓ s and ✗s according to the rule just stated. One then marks parent nodes as factual when all their children have been so marked, and as anti-factual when at least one their children is anti-factual. P is grounded-true, on this way of doing it, iff some decorated dependence tree has $|P|^{\mathbf{t}}$✓ at the top.

Here for instance is an undecorated tree for *Something Russell believed was true, and something he believed was false*, on the hypothesis that Russell believed (at least) that *Ice is cold* (I), which is true, and that *Jello is hot* (J), which is false.

$$|(\exists x,y{:}\ Brx \,\&\, Bry)\ T(x) \,\&\, F(y)|^{\mathbf{t}}$$
$$|T(I)|^{\mathbf{t}} \qquad |F(J)|^{\mathbf{t}}$$
$$|I|^{\mathbf{t}} \qquad |J|^{\mathbf{f}}$$

To decorate it, we start by appending ✓ to any terminal node that is factual. As it happens they both are, so we have two ✓s to tack on.

$$|(\exists x,y{:}\ Brx \,\&\, Bry)\ T(x) \,\&\, F(y)|^{\mathbf{t}}$$
$$|T(I)|^{\mathbf{t}} \qquad |F(J)|^{\mathbf{t}}$$
$$|I|^{\mathbf{t}}\checkmark \qquad |J|^{\mathbf{f}}\checkmark$$

That was stage 1 of the operation. Now we move gradually upward, checking off at stage $n+1$ any nodes each of whose children were checked off at stage n. This yields, at stage 2,

$$|(\exists x,y{:}\ Brx \,\&\, Bry)\ T(x) \,\&\, F(y)|^{\mathbf{t}}$$
$$|T(I)|^{\mathbf{t}}\checkmark \qquad |F(J)|^{\mathbf{t}}\checkmark$$
$$|I|^{\mathbf{f}}\checkmark \qquad |J|^{\mathbf{f}}\checkmark$$

and at stage 3,

$$|\,(\exists x,y{:}\ Brx \,\&\, Bry)\ T(x) \,\&\, F(y)|^{\mathbf{t}}\checkmark$$
$$|T(I)|^{\mathbf{t}}\checkmark \qquad |F(J)|^{\mathbf{t}}\checkmark$$
$$|I|^{\mathbf{t}}\checkmark \qquad |J|^{\mathbf{f}}\checkmark$$

A decorated tree headed by $|\varphi|^{\mathbf{t}}\checkmark$ means that φ is grounded-true. So, *Not everything Russell said was true, nor was it all false* is true by the lights of Kripke's grounding semantics.

Now let's try the rules out on some trickier examples, starting with the Liar L (= $\neg T(L)$), the Truth-Teller, and so on.

$$|L|^{\mathbf{t}}\ (=\ |\neg T(L)|^{\mathbf{t}}) \qquad\qquad |L|^{\mathbf{f}}\ (= |\neg T(L)|^{\mathbf{f}})$$
$$|T(L)|^{\mathbf{f}} \qquad\qquad |T(L)|^{\mathbf{t}}$$
$$|L|^{\mathbf{f}} \qquad\qquad |L|^{\mathbf{t}}$$
$$|T(L)|^{\mathbf{t}} \qquad\qquad |T(L)|^{\mathbf{f}}$$
$$|L|^{\mathbf{t}} \qquad\qquad |L|^{\mathbf{f}}$$
$$\text{etc} \qquad\qquad \text{etc}$$

(1) Liar

That neither tree terminates means that L is neither grounded-true nor grounded-false. Attempts to decorate either one never get off the ground since there are no terminal nodes to start from. Note that the Liar trees not only conflict with each other (that's by design) but also each with itself; each contains $|P|^{\mathbf{t}}$ and $|P|^{\mathbf{f}}$ for the same sentence P.

The Truth-Teller K (= $T(K)$) again has two trees, each with a single infinite branch. The difference is that K's trees are, taken individually, consistent; neither assigns truth and falsity to any sentence P. There is to that extent a consistent scenario where K is true, and another where K is false. Still, that neither tree terminates means that K is ungrounded, that is, neither grounded-true nor grounded-false.

(2) Truth-Teller

Now the truthfulness-teller H. Assume that Moore's other statements (other than H) are I = *Ice is cold* and $\neg J$ = *Jello isn't hot*; then $H = T(I)\ \&\ T(\neg J)\ \&\ T(H)$. The trees of interest are

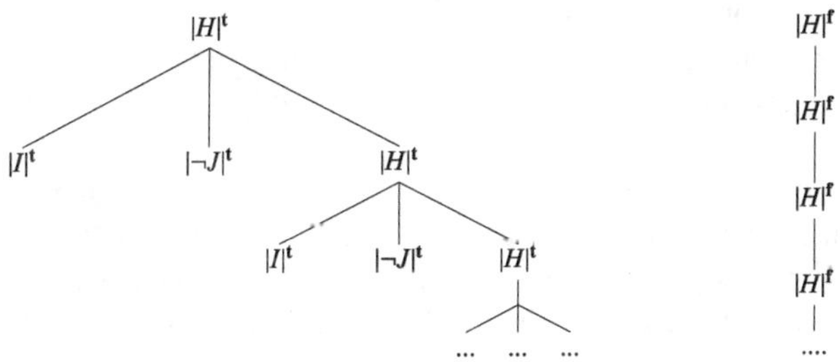

(3) Truthfulness-Teller

From the right-hand tree we see that H is not grounded-false. The tree for $|H|^{\mathbf{t}}$ has an infinite branch too, though, so H is not grounded-true either. Both of the trees are consistent, as with K. Officially then, H is underdetermined, just like the Truth-Teller. But that is not how it strikes us. It strikes us as true, or something very like true. There might be some support for this idea in the fact that $|H|^{\mathbf{t}}$'s tree is "better"—more grounded in nonsemantic facts—than $|H|^{\mathbf{f}}$'s. We'll return to this theme in a moment.

7.6 Immodesty

The Truthfulness-Teller is ungrounded, on Kripke's theory, because it immodestly extends to itself the compliment (truth) that it pays to other sentences. I can think of two ways to make it less immodest, so as to give it a better shot at truth. We could muck with the subject term, so that it covered fewer sentences. Or we could scale back the predicate, so that it attributed a weaker property.

On the first strategy, we mistake *All my statements are true* for H_1, which attributes truth only to Moore's *other* statements. That we were taking it for H_1 nicely explains why H would strike us as true. H_1 really *is* true; Moore's other statements really do have the property (truth) that's attributed to them. This approach also explains why the Truth-Teller seems worse off than the Truthfulness-Teller. If we cut back K's subject term ("this very sentence"), then nothing is left; there are no other statements that K describes inter alia as true. K is worse off than H because there is no worthwhile K_1 standing to it as H_1 stands to H.

These results are obtained, however, by twisting H's intuitive content out of recognition. *All my statements are true...with the possible exception of this one* is the statement of some kind of trickster, not a George Edward Moore. To exempt his declaration of honesty from its own extension is the last thing Moore wants. Here then is our first condition on a satisfactory solution: *All my statements are true* should not make an exception of itself.

Doesn't this make the problem unsolvable, though? For Moore's declaration not to make an exception of itself would seem to mean that it is one of the statements that it describes as true. But then it has a Truth-Teller inside it, with the truth-destroying ungroundedness that that entails.

But there's a second thing we could try—targeting not the subject term but the predicate. Perhaps what Moore meant is H_2: Everything I say is true-to-the-extent-evaluable.

This again does violence to the content. Suppose Moore had on other occasions uttered a bunch of ungrounded nonsense: Liars and Truth-Tellers and whatever other semantic pathologies you like. His statements are true to the extent evaluable, just because they are not evaluable. The Truthfulness-Teller is not so easily saved. If I say, *All my statements are true*, when in fact NONE have this property, my claim may be many things, but "true" is not one of them. *All my statements are true* should attribute *truth*, not something weaker like truth-where-evaluable. And now we are back in trouble, because if H calls itself true, then it is NOT true, on account of being ungrounded; to be true, it must have been true already.

What other way of modifying the Truthfulness-Teller is there, though, if we are not allowed to make the subject term more demanding, or the predicate less so? *Maybe it is not H that needs to be modified, but the claim we make on its behalf.* Rather than calling it true, period, perhaps we can call it true about a certain subject matter: the facts, as it might be. This is what we suggest below (Sect. **7.11**): even if H is not true full stop, still it is true to the facts. The problem of course is to identify this new subject matter. I propose to creep up on it slowly, by way of liberalized dependence trees.

7.7 TRUTH and Grounding

For Kripke, in the first instance anyway, a sentence is true (false) only if it's *grounded*-true (-false). The Truthfulness-Teller seems to cast doubt on this idea. Let's remind ourselves of what it means for a sentence to be grounded-true.

0. P is true iff $|P|^{\mathbf{t}}$ sits atop a dependence tree all of whose branches terminate in facts.

It is grounded-false iff $|P|^{\mathbf{f}}$ sits atop a dependence tree all of whose branches terminate in a fact. (Or, what comes to the same, $\neg P$ is grounded-true.) Could the lesson of H be that grounding is too strict a condition? As a first stab at something looser, consider

1. P is TRUE (first stab) iff $|P|^{\mathbf{t}}$ sits at the top of a dependence tree
(i) *some* of whose branches terminate in facts, and
(ii) all of whose terminating branches terminate in facts.

(I write TRUE so as not to beg any questions about the identity of this truth-like property with the one Kripke is attempting to analyze.) A sentence is TRUE, in other words, if $|P|^{\mathbf{t}}$ ✓ heads a decorated dependence tree constructed to slightly weaker specifications: a parent node is marked ✓ iff (i) some of its children are marked ✓, and (ii) none if its (other) children are marked ✗. The earlier requirement was that a parent node is validated iff *all* its children are validated.

This makes the Truthfulness-Teller TRUE, which is good, but it also makes the Truth-Teller TRUE, or at least treats it that way in certain constructions. An example is $I\&\neg J\&K$, where I and $\neg J$ are plain truths, and K is again the Truth-Teller. The tree is

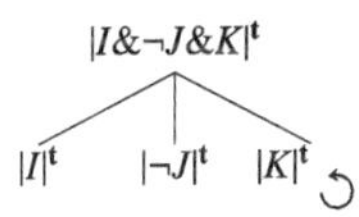

(4) Truth-Teller Plus

Note, the ↺ notation is to indicate that a node depends on itself; the tree fully spelled out puts $|K|^{\mathbf{t}}$ on top of an infinite descending chain of $|K|^{\mathbf{t}}$'s. $I\&\neg J\&K$ meets the condition [**1.**] lays down for TRUTH: some branches terminate in facts, the others don't terminate. This seems just wrong, however. How can $I\&\neg J\&K$ be TRUE, if K, its third conjunct, lacks this property?

7.8 TRUTH and Tethering

I want to go back now to an idea from Sect. **7.7**: some ungrounded attributions are closer to being grounded than others. A glance at their trees makes clear that $|H|^{\{\mathbf{t}}$, for instance, is less ungrounded than $|I\&\neg J\& K|^{\mathbf{t}}$, which is less ungrounded than $|K|^{\mathbf{t}}$, and also less ungrounded than $|H|^{\mathbf{f}}$. In what sense, though?

A node is *tethered*, let us say, if it has a finite path to the facts—a fact, recall, is a non-semantic atomic attribution $|A|^{\mathbf{t}}$ ($|A|^{\mathbf{f}}$) such that A is true (false) in the underlying model. A branch or tree is tethered if all its nodes are. Looking back now at the trees provided for $|H|^{\mathbf{t}}$ and $|H|^{\mathbf{f}}$, we see that they greatly differ in this respect. In the first, every node is tethered; every node has a finite path to the facts. In the second, *no* node has this property. Maybe the requirement ought to be, not no *infinite* branches, but no *untethered* branches, where a branch is tethered iff every node is tethered; every node has a (finite) path to the facts.

2. P is TRUE (second stab) iff $|P|^{\mathbf{t}}$ has a tethered dependence tree.

The Truthfulness-Teller is TRUE by this strengthened standard too. Each occurrence of $|H|^{\mathbf{t}}$ has *two* paths to the ground, ending in $|I|^{\mathbf{t}}$ and $|\neg J|^{\mathbf{t}}$ respectively. K's conjunction with I and $\neg J$ is not TRUE according to [**2.**], since the tree has a branch $|K|^{\mathbf{t}} \rightarrow |K|^{\mathbf{t}} \rightarrow |K|^{\mathbf{t}} \rightarrow$ ... all of whose elements are untethered.

This idea of tethering speaks to the "comparative" problem of how H can be better off than K, even though it in some sense includes K, or an analogue of K. H's advantage is that every last bit of it hooks up with the facts— every node on its tree depends on them—whereas K is floating around absolutely untethered, depending only on itself.[4]

A problem emerges when we consider the *Un*truthfulness or Mendacity-Teller, *Everything I say is false* (henceforth M). Suppose that my only other statement is J (*Jello is hot*), which is false. Then $M = F(J)\&F(M)$; whence $|M|^{\mathbf{t}}$ has the following as one of its trees.

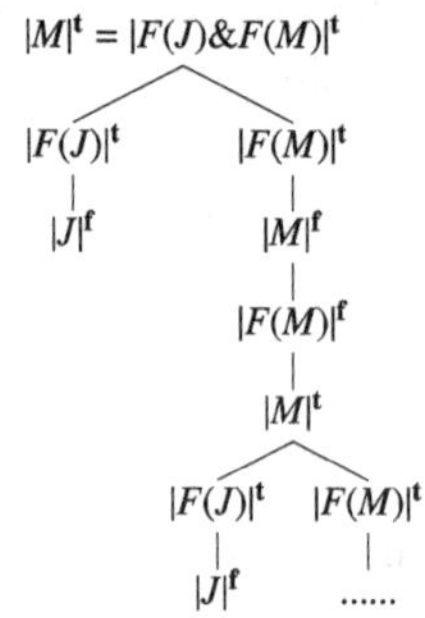

(5) Mendacity-Teller

Every node here has a finite path to $|J|^{\mathbf{f}}$; $|J|^{\mathbf{f}}$ is factual; so every node here is tethered. "Everything I say is false" ought, then, according to [**3.**], to be TRUE. But it is in reality paradoxical, since if M is true, then, given that it has $F(M)$ as a conjunct, it is FALSE. (Whereupon it is TRUE after all, and so on.)

Notice something objectionable about tree (5), however; it has $|M|^{\mathbf{t}}$ on top and $|M|^{\mathbf{f}}$ further down, making the tree as a whole inconsistent. Perhaps

3. P is TRUE (third try) iff $|P|^{\mathbf{t}}$ has a *consistent* tethered dependence tree.

This is better, but even a consistent tethered tree is not enough, as we see from an example of Vann McGee's. Let N_1 be N_2 *is false and ice is cold*, while N_2 is N_1 *is false and ice is cold*. Surely N_1 cannot be TRUE, for then N_2 would have to be FALSE, which is ruled out by symmetry considerations; there is no reason why N_2

[4]There could be an "unwinding" of K that does not depend on itself, yet is equally untethered. Kripke notes the possibility of "an infinite sequence of sentences P_i, where P_i says that P_{i+1} is true" (Kripke 1975, 693). For unwindings more generally see Schlenker (2007) and Cook (2014).

should be the FALSE one rather than N_1. Yet here is a consistent tethered tree for $|N_1|^{\mathbf{t}}$[5]:

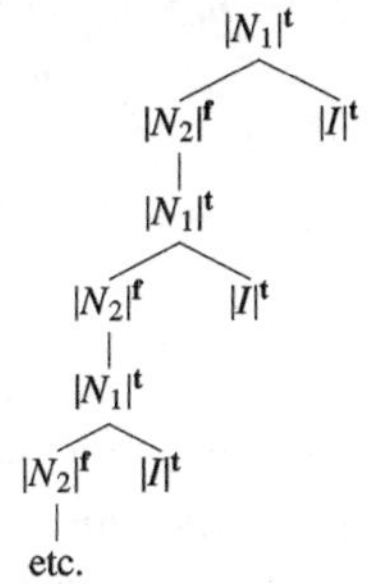

(6) McGee Tree

What is interesting is that such a tree is *also* constructible for $|N_1|^{\mathbf{f}}$; it mirrors the tree for $|N_2|^{\mathbf{f}}$ that is embedded in tree (6).

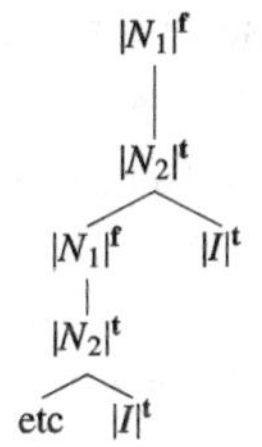

(7) McGee's Other Tree

The McGee trees show that [**3.**] needs to be tightened up a bit:

4. P is TRUE (fourth and final stab) iff $|P|^{\mathbf{t}}$ has, while $|P|^{\mathbf{f}}$ lacks, a consistent tethered dependence tree.

The Truthfulness-Teller H is TRUE, according to [**4.**], given that $|H|^{\mathbf{t}}$ has a tethered tree, if no consistent tethered tree can be constructed for $|H|^{\mathbf{f}}$. The only possible tree for $|H|^{\mathbf{f}}$, assuming as usual that Moore's other statements are I and $\neg J$ (both true), is

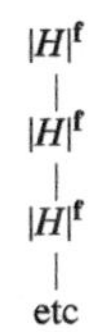

(8) Truthfulness-Teller False[6]

[5]Compressed for readability.

This[6] again is untethered, containing not even one node with a finite route to the ground. $|H|^{\mathbf{t}}$ is thus the only one of $|H|^{\mathbf{t}}$, $|H|^{\mathbf{f}}$, to have a consistent tethered tree, which justifies our preference for *I (Moore) never lie* over *I (Moore) do sometimes lie.*

7.9 Fixed Points

Subject matters as we are going to be conceiving them (following Lewis) are equivalence relations on worlds. What plays the world role in this application are fixed points. These are much better known than dependence trees, but require a bit of explanation as they haven't been mentioned yet in this paper.

A sentence is grounded-true (-false), we said, iff the corresponding attribution $|S|^{\mathbf{t}}$ ($|S|^{\mathbf{f}}$) has a dependence tree all of whose branches terminate in facts—atomic attributions $|A|^{\mathbf{t}}$ ($|A|^{\mathbf{f}}$) such that A really is true (false) in the underlying model. Kripke's definition is different; he uses not trees but sets of attributions satisfying certain closure conditions. A *fixed point* is a consistent set of attributions $\mathcal{P}$ such that

(A) if A is atomic, $|A|^{\mathbf{t}} \in \mathcal{P}$ ($|A|^{\mathbf{f}} \in \mathcal{P}$) iff A is true (false) in the underlying model.[7]
(N) $|\neg S|^{\mathbf{t}} \in \mathcal{P}$ iff $|S|^{\mathbf{f}} \in \mathcal{P}$; $|\neg S|^{\mathbf{f}} \in \mathcal{P}$ iff $|S|^{\mathbf{t}} \in \mathcal{P}$
(D) $|S \vee S'|^{\mathbf{t}} \in \mathcal{P}$ iff $|S|^{\mathbf{t}} \in \mathcal{P}$ or $|S'|^{\mathbf{t}} \in \mathcal{P}$; $|S \vee S'|^{\mathbf{f}} \in \mathcal{P}$ iff $|S|^{\mathbf{f}} \in \mathcal{P}$ and $|S'|^{\mathbf{f}} \in \mathcal{P}$
(U) $|\forall x\ \varphi(x)|^{\mathbf{t}} \in \mathcal{P}$ iff $|\varphi(n)|^{\mathbf{t}} \in \mathcal{P}$ for each n; $|\forall x\ \varphi(x)|^{\mathbf{f}} \in \mathcal{P}$ iff $|\varphi(n)|^{\mathbf{f}} \in \mathcal{P}$ for some n
(T) $|T(S)|^{\mathbf{t}} \in \mathcal{P}$ iff $|S|^{\mathbf{t}} \in \mathcal{P}$; $|T(S)|^{\mathbf{f}} \in \mathcal{P}$ iff $|S|^{\mathbf{f}} \in \mathcal{P}$

If these rules look familiar, and they should, it's because the left-to-right directions of **(A)**-**(T)** are the same as the tree rules **(a)**-**(t)** laid down in Sect. **7.5**.

Suppose that $\mathcal{A}$ is a set of nonsemantic atomic attributions. $\mathcal{P}$ is a fixed point over $\mathcal{A}$ iff it is a fixed point whose nonsemantic atomic attributions are precisely those in $\mathcal{A}$. Kripke finds work for lots of fixed points, but the one he particularly emphasizes is

$$\mathcal{G}_{\mathcal{A}} = \text{the least fixed point over } \mathcal{A}.$$

A sentence S is grounded-true (-false), for Kripke, given nonsemantic facts $\mathcal{A}$, iff $|S|^{\mathbf{t}}$ ($|S|^{\mathbf{f}}$) belongs to $\mathcal{G}_{\mathcal{A}}$—or, what is really no different, $|S|^{\mathbf{t}}$ ($|S|^{\mathbf{f}}$) belongs to *every* fixed point over $\mathcal{A}$. This conforms to our tree-based definition, since $\mathcal{G}_{\mathcal{A}}$ turns out (unsurprisingly) to be precisely the set of attributions with trees that terminate in the facts, as represented by $\mathcal{A}$.[8]

A prima facie advantage of fixed points over trees is that they make for a richer taxonomy. P is *paradoxical*, for instance, if no fixed point contains either $|P|^{\mathbf{t}}$ or $|P|^{\mathbf{f}}$.[9] It is *unstable* iff it is true in some consistent fixed points and false in others. It

[6] Taken from (3) above.
[7] The underlying model M is a model, possibly partial, of the T-free part of the language.
[8] Yablo (1982).
[9] No consistent fixed point that is; but we have defined fixed points so that all of them are consistent.

is *stable* iff it is true in some fixed points and false in none (or vice versa); it receives in other words the same truth-value in every fixed point that's defined on it, and there are some. P is *intrinsically* true iff it is true in a thoroughly stable fixed point, meaning, one defined only on stable sentences.

Both H and K—the Truthfulness Teller and the Truth Teller—are unstable; they are true in some fixed points, clearly, and false in others. $T(K)\supset T(K)$, however, is stably true: true in those consistent fixed points where it has a truth-value at all. Is it intrinsically true? No, for $T(K)\supset T(K)$ is evaluable only in fixed points that assign a value to K, and K is unstable. An intrinsic truth in the same neighborhood is $E = \neg(T(E)\&\neg T(E))$—"This very sentence is not both true and untrue." Its one potential truth-value is *true*, and the one unchanging basis for that truth-value is the fact just mentioned, the fact of E'S truth. The intrinsic attributions can be joined together into a single compendious fixed point $\mathcal{I}$, Kripke shows, the "maximal intrinsic fixed point." P is intrinsically true (false) just if it is true (false) in $\mathcal{I}$.

Now, if one is looking for a compliment that can be paid to ungrounded sentences—which we are, given the true-seemingness of the Truthfulness-Teller—intrinsic truth is a Kripkean's first thought. ("This sentence is not both true and false" is intrinsically true, as just noted.) It's a compliment that cannot be paid to H, however. The Truthfulness-Teller patterns with the Truth Teller in being not even stably true, much less intrinsically so. If we stipulate that H is false in a fixed point, we then provide a reason for its falsity; it is a counterexample to the generalization that everything Moore says is true. If we stipulate that it is true, we eliminate the one possible counterexample to its truth, namely, itself.

7.10 TRUTH in Fixed Points

Can *This speaker is truthful* really be no better than *This sentence is true*, from a fixed point perspective? That would be surprising, given the close connection between fixed points and trees.

If we gather together all the attributions $|\varphi|^{v}$ on a consistent tree, we get a partial valuation $\mathcal{V}$ that is closed under the left-to-right directions of (**A**)-(**T**); a valuation like that is called *sound*. Sound valuations generate fixed points $\mathcal{V}^*$ under repeated application of the right-to-left directions of (**A**)-(**T**). Every tree is in that sense the seed of a fixed point. And of course there will be other fixed points above $\mathcal{V}^*$, involving attributions not forced by $\mathcal{V}$, but allowed by it.

This forced/allowed distinction is the key to distinguishing H from K in fixed point terms. K has no factual prerequisites and faces no factual threats. No matter what the ground-level facts $\mathcal{A}$ may be, K is true in some fixed points above $\mathcal{A}$ and false in others. K and $\neg K$ are both *unconditionally possible*: each holds in *some* fixed point above every factual ground.

The Truthfulness-Teller is different in this respect. H can be true only in fixed points making Moore's other statements true: ice has got to be cold and Jello cannot be hot. H is only *conditionally* possible. The result $\neg H$ of negating it is, however, unconditionally possible just like K; whatever the ground-level facts may be, we can consistently treat H as false by virtue of its own falsity. H is more beholden to the actual facts than its negation, and than K and its negation. Of the four, it is the only one that owes its construability as true to the way things actually turned out.

Now this is not quite enough for TRUTH, for it holds of *K&I*—*This sentence (up to the ampersand) is true & Snow is white*—as well that (i) it owes its construability as true to the way things turned out, while (ii) its negation is construable as true no matter what (by letting K be false). And yet *K&I* certainly does not strike us as TRUE, to repeat an observation made earlier.

Suppose we use *fact-dependent* for the property of being construable as true in *these* factual circumstances—$\mathcal{A}_@$—but not in *all* factual circumstances. The problem with *K&I* is that while it is fact-dependent taken as a whole, its first conjunct is unconditionally possible or fact-free. What is special about the Truthfulness-Teller is that it is *thoroughly* fact-dependent, not an amalgam of something fact-dependent with something fact-free.

How to define this in fixed point terms? Consider the fixed points above $\mathcal{A}_@$. For one of these to be fact-dependent, all of its component attributions should be fact-dependent; it should contain nothing that is unconditionally possible, nothing that is construable as true no matter what. An attribution is *thoroughly fact-dependent* iff it belongs to a fixed point *all* of whose attributions are fact-dependent,

This is reminiscent of what we said about tethered trees; the attributions on them may not all be grounded, but they all have finite paths to the ground. The two notions—tethered tree and thoroughly fact-dependent fixed point—are connected, it turns out. $|\varphi|^t$ heads a consistent tethered tree just if φ is true in at least one thoroughly fact-dependent fixed point. (For short, one fact-dependent fixed point.)

If a tree is untethered, it has a node n with no finite path to the non-semantic atomic facts. The subtree that n heads must therefore be free of such facts. Let $\mathcal{N}$ be the subtree's contents = the set of all attributions on it. These attributions form a sound set (the contents of any tree make a sound set) that is consistent with any $\mathcal{A}$ (because $\mathcal{A}$ is made up of ground-level attributions and $\mathcal{N}$ is free of such attributions). $\mathcal{N} \cup \mathcal{A}$ generates a fixed point containing the attribution in n ($|\varphi|^{\mathrm{v}}$, let's say) by application of the right-to-left directions of closure rules (**A**)-(**T**). $|\varphi|^{\mathrm{v}}$ is fact-independent because $\mathcal{A}$ was arbitrary. *An untethered tree must therefore contain elements that are fact-independent.*

Suppose conversely that an attribution $|\varphi|^{\mathrm{v}}$ is fact-independent, that is, $|\varphi|^{\mathrm{v}}$ is unconditionally possible. Then for every $\mathcal{A}$ whatsoever there is a fixed point above $\mathcal{A}$ that assigns ^{v} to φ. This is so in particular if $\mathcal{A}$ is the empty set. Fixed points by definition satisfy conditions (**N**) for negation, (**D**) for disjunction, (**U**) for quantification, and (**T**) for truth. The left-to-right directions of these rules give us all we need to construct a tree for $|\varphi|^{\mathrm{v}}$. The tree is going to be untethered because there were no ground-level attributions in the fixed point: $\mathcal{A}$ is the empty set. We have shown that

Lemma $|\varphi|^{\mathrm{v}}$ *has a consistent tethered dependence tree iff it belongs to a fact-dependent fixed point.*

From this it follows that

Theorem *φ is TRUE iff it is true in at least one fact-dependent fixed point and false in no such fixed points.*[10]

The theorem bears on a problem posed above: can a subject matter be identified such that φ is TRUE iff it is true about that subject matter?

7.11 True to the FACTS

A *subject matter*, for Lewis, is an equivalence relation on worlds.[11] Sentence S is *wholly about subject matter* M just if S's truth-value never varies between M-equivalent worlds. *The number of stars is prime* is wholly about how many stars there are, since worlds with equally many stars cannot disagree on whether their stars are prime in number.[12] *The number of stars exceeds the number of planets* is *not* wholly about the number of stars, since its truth-value can change though the number of stars holds fixed. Now we define the notion of truth at a world where a given subject matter is concerned:

(TAM) S is true about M in a world w iff it is true in a world M-equivalent to w.

Worlds for these purposes can be fixed points, as indicated earlier.[13] The facts in two worlds are the same if, although they may evaluate T-sentences differently, regular old non-semantic atomic sentences have the same truth-value in both of them. Suppose that w and w' are fact-dependent. They agree on subject matter F, short for FACTS, just if the same facts obtain in both of them.

(SMF) Worlds are F-equivalent iff
(i) both are fact-dependent, and
(ii) the same facts obtain in each.

By (TAM), φ is true about F in w iff φ it is true in a fact-dependent fixed point w' agreeing with w in its non-semantic atomic facts. Consider now truth about F in the actual world $w_@$, defined as the least fixed point based on the actual facts.
Theorem* *φ is TRUE iff it is true in the actual world where the FACTS are concerned.*

This is a shorter way of saying, as we did above, that to be TRUE is to be true in at least one fact-dependent fixed point whose facts are the actual ones. The paper could end right here, but I have a parting speculation I'd like to get on the table.

The notion of aboutness we get from Lewis is important and interesting. But it is not the only one possible. We saw for instance that *The number of stars exceeds*

---

[10]If φ is TRUE, then $|\varphi|^{\mathbf{t}}$ has a consistent tethered dependence tree and $|\varphi|^{\mathbf{f}}$ doesn't. By the Lemma, φ is true in a fact-dependent fixed point but not false in any fact-dependent fixed points. The converse is similar.

[11]Lewis (1988).

[12]Lewis (1988).

[13]We will be interested only in fact-dependent fixed points, more carefully, fixed points that are fact-dependent relative to some choice $\mathcal{A}$ of non-semantic atomic facts.

the number of planets is not in Lewis's sense about the number of stars, since its truth-value can change though the number of stars remains what it is. But there is another sense in which *The number of stars exceeds the number of planets* IS about the number of stars; its truth-value is *sensitive* to how many stars there are; there can't be zero stars, for instance, compatibly with the stars outnumbering the planets.

One can imagine conversely a sentence that is about the number of stars in the supervenience sense, but not the sensitivity or difference-making sense. *The number of stars is positive* is supervenience-about how many stars there are, in that worlds M-alike are always **S**-alike, but not differentially about how many stars there are, in that M-different worlds—worlds with unequally many stars—do not thereby differ in whether the number of stars in them is positive (Yablo 2014).

Where Lewis's *supervenience*-based notion of aboutness focuses on whether φ's semantic properties hold fixed when you hold the state of things wrt M fixed, there is another notion, the *differential* notion, that looks rather at how φ's semantic properties are apt to *change* if you vary the state of things with respect to M. H and K may be equally about the facts (or not) in the supervenience sense, but they are not equally about the facts in the difference-making sense. What do I mean by this?

Changing the facts $\mathcal{A}$ has no effect on the Truth Teller whatever—it can be true or false as you please—but the Truthfulness Teller loses its shot at truth if we move to a world where Jello is hot. The Truthfulness Teller outdoes the Truth Teller differential-aboutness-wise because *changing* the facts has the potential to *change* H's semantic properties, but not the semantic properties of K. This links up with our talk earlier of fact-dependence and fact-freedom, and it might be possible to reformulate those earlier notions in differential aboutness terms.

7.12 Conclusion

The way is now clear for Moore to call himself honest without falling afoul of the strictures imposed by the best known theory of truth. Knights are encouraged to avail themselves of this opportunity, too.

References

Barwise, J., & Etchemendy, J. (1989). *The liar: An essay on truth and circularity*. USA: Oxford University Press.

Boolos, G. (1971). The iterative conception of set. *The Journal of Philosophy,* 215–231.

Cook, R. T. (2014). *The Yablo paradox: An essay on circularity*. Oxford: Oxford University Press.

Davis, L. (1979). An alternate formulation of Kripke's theory of truth. *Journal of Philosophical Logic*, *8*(1), 289–296.

Hazen, A. (1981). Davis's formulation of Kripke's theory of truth: A correction. *Journal of Philosophical Logic*, *10*(3), 309–311.

Herzberger, H. G. (1970). Paradoxes of grounding in semantics. *The Journal of Philosophy*, 145–167.
Kripke, S. (1975). Outline of a theory of truth. *Journal of Philosophy*, *72*, 690–716.
Lewis, D. (1988). Statements partly about observation. In *Papers in philosophical logic*. Cambridge: Cambridge University Press.
McLarty, C. (1993). Anti-Foundation and Self-Reference. *Journal of Philosophical Logic*, *22*(1), 19–28.
Mirimanoff, D. (1917). *Les antinomies de Russell et de Burali-Forti: et le problème fondamental de la théorie des ensembles*. Enseignement mathématique.
Schlenker, P. (2007). The elimination of self-reference: Generalized yablo-series and the theory of truth. *Journal of philosophical logic*, *36*(3), 251–307.
Smullyan, R. (1986). *What is the name of this book?* Touchstone Books.
Yablo, S. (1982). Grounding, dependence, and paradox. *Journal of Philosophical Logic*, *11*, 117–138.
Yablo, S. (1993). Hop, skip and jump: The agonistic conception of truth. *Philosophical Perspectives*, *7*, 371–396.
Yablo, S. (2006). Circularity and paradox. In T. Bolander, V. F. Hendricks, & S. A. Pedersen (Eds.), *Self-reference* (pp. 139–157). Stanford: CSLI Publications.
Yablo, S. (2014). *Aboutness*. Princeton: Princeton University Press.
Yuting, S. (1953). Paradox of the class of all grounded classes. *Journal of Symbolic Logic*, *18*(2), 114.

Chapter 8
What I Tell You Three Times Is True

Martin Davis

Abstract A very brief look at self reference in literature and art.

Keywords Riemann hypothesis · Self reference in art
Self reference in literature · Undecidability

> *"Just the place for a Snark!" the Bellman cried,*
> *As he landed his crew with care;*
> *Supporting each man on the top of the tide*
> *By a finger entwined in his hair.*
>
> *Just the place for a Snark! I have said it twice:*
> *That alone should encourage the crew.*
> *Just the place for a Snark! I have said it thrice:*
> *What I tell you three times is true.*
>
> —Lewis Carroll's "The Hunting of the Snark"

When the Bellman refers to his previous utterances, he is in fact referring to previous lines in this delicious piece of nonsense. Lewis Carroll knew very well that employing this bit of self-reference would add a delightful tickle to the absurd image of a crew member being supported by a finger in his hair, while also calling attention to the equal absurdity of foolish speakers who imagine that repetition will enhance their credibility.

Raymond Smullyan knows very well how adding such a tickle can enhance the pleasure of his readers. In his famous puzzle book *This Book Needs No Title* the self reference in the title is also present in many of his puzzles Smullyan (1986). And some of the same reasoning one finds in these puzzles is used to startling effect in his serious treatise Smullyan (1994) where they lead to dizzying heights of abstraction and then back down to earth in myriads of applications including what amounts to a textbook on combinatory logic.

Of course Raymond is not alone. The use of self reference by writers has been so ubiquitous that academics discuss it in a context they call metafiction. One of

M. Davis (✉)
New York University and University of California, Berkeley, USA
e-mail: martin@eipye.com

M. Fitting and B. Rayman (eds.), *Raymond Smullyan on Self Reference*,
Outstanding Contributions to Logic 14, https://doi.org/10.1007/978-3-319-68732-2_8

Fig. 8.1 René Magritte, *Ceci N'est Pas Une Pipe* ©2016 C. Herscovici/Artists Rights Society (ARS) New York

my own favorite examples is in Laurence Sterne's 18th century *Tristram Shandy*. Mr. Shandy, "gentleman", tells us:

> …my life and opinions are likely to make some noise in the world, and, …will take in all ranks, professions, and denominations …– be no less read than the *Pilgrim's Progress* itself …

We are informed at the outset that Tristram was convinced that he was in some unspecified way defective, and that this defect would have been avoided had his parents "minded what they were about when they begot [him]". Instead, his father was distracted while in the midst of this act, by his wife's inquiring, "Pray, my dear, have you not forgot to wind up the clock?" thus interfering with the necessary transfer of "animal spirits".[1]

Self reference can even be found in epic tales that were sung by bards long before they were written down. In Homer's Odyssey, we meet the bard Demodocus, who, like Homer himself, though blind, had a sweet voice. Demodocus sings of a quarrel between Achilles and Odysseus before Troy.

Self reference in visual works of art are perhaps more directly striking than those in literary works. Perhaps the most famous example is Magritte's *Ceci n'est pas une pipe*"[2] in which this very title is painted in handwriting directly on the canvas below a painted image of a pipe, Fig. 8.1. This painting has been much parodied and imitated. My personal favorite example of this is *This is Not a Slide* by Virginia Davis (my wife of over six decades), Fig. 8.2. When she painted this she was in Paris in residence at the Cité Internationale des Arts. It is painted on a linen canvas that she had woven in her studio in New York using threads some of which had had color applied before weaving. At the time artists were still using 35 mm slides for photographs of their work and she painted the image of just such a slide on the canvas. Below that image she painted the words "Ceci n'est pas une diapositive", *diapositive* being the French word for such a slide.

Another interesting example by Magritte is his *La tentative de l'impossible*[3] in which the painter is shown with his palette and brushes painting in the left arm

---

[1] Sterne (2003), pp. 5–6, 8.

[2] i.e., This is not a pipe.

[3] i.e., Attempt of the impossible.

Fig. 8.2 Virginia Davis, *This Is Not A Slide* ©2003 Virginia Davis

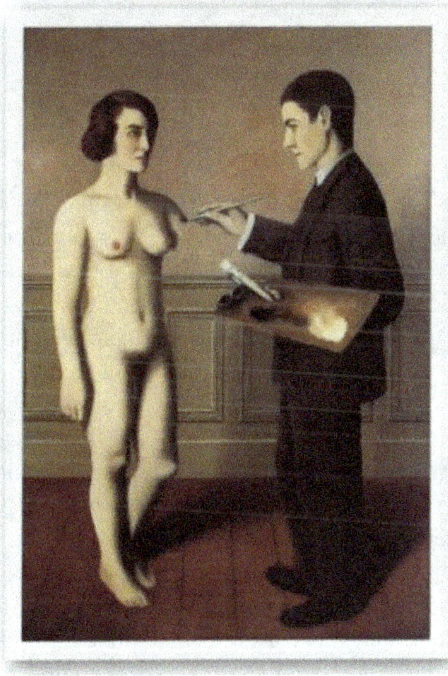

Fig. 8.3 René Magritte, *La tentative de l'impossible* ©2016 C. Herscovici/Artists Rights Society (ARS) New York

of an otherwise complete naked woman standing before him, Fig. 8.3. A similar impossibility (one that I think would appeal particularly to Raymond) is an amazing drawing by Saul Steinberg, Fig. 8.4, of multiple images of himself drawing himself in a recursive loop.[4] Probably the most famous example of an artist achieving self reference by including himself in the act of making the painting is *Las Meninas* by Diego Velazquez, Fig. 8.5. A careful look at this intricate composition will find the painter painting the scene from an adjoining room.

---

[4]Thanks to Judith Dunford who called my attention to this example.

Fig. 8.4 *Untitled*, Saul Steinberg, *Untitled*, 1944 Ink on paper, $19\frac{1}{2} \times 14\frac{1}{2}$ in. Beinecke Rare Book and Manuscript Library, Yale University ©The Saul Steinberg Foundation / Artists Rights Society (ARS), New York

The examples including text as part of the painting have the interesting property that their point would be lost on a viewer ignorant of the language in which the text is written. In the case of the Virginia Davis painting, the viewer would need some understanding of both English and French. I find it interesting that something similar is the case for Gödel's undecidable sentence. Someone who doesn't understand the code by means of which metamathematical concepts are translated to formulas of the system will be totally unaware of the fact that the sentence can be seen as expressing its own unprovability. It is certainly conceivable that some problems that have seemed intractable do express their own unprovability, but with respect to a code of whose very existence we remain totally unaware. For example, to speculate wildly, perhaps there is an arithmetic equivalent of the Riemann Hypothesis that expresses its own unprovability in Peano Arithmetic with respect to some hidden code.

It has been over eight decades since Gödel announced his incompleteness theorem, and much has been written about it, a good deal of which shows no real understanding of the actual content of the theorem. But Gödel's work has had little if any effect on mathematical practice. This is in contrast to the closely related matter of unsolvability in the sense of the non-existence of algorithms for the solution of various problems. Although in this case the initial unsolvability results also made use of self reference and diagonalization, by using the technique of reducing problems of evident mathematical interest to one of those original problems, much of obvious mathematical interest has been accomplished. In particular, in the instance of word problems and of Diophantine equations, problems that had been previously posed as seeking a positive solution have been proved unsolvable. It is easy to speculate that some of the open questions with which number theory abounds are undecidable in Peano Arithmetic. In fact, in his famous Gibbs lecture, Gödel did suggest exactly

Fig. 8.5 *Las Meninas*, Diego Velazquez [Public Domain]

that with respect to the Riemann Hypothesis.[5] But of course all of this remains a matter of speculation.

Harvey Friedman has been a zealous worker in this area, producing a large number of interesting Π^0_1 and Π^0_2 combinatorial propositions that are unprovable in ZFC but that become provable when a suitable large-cardinal axiom is added.[6] But as striking as these results are, they do not seem to provide a path toward showing anything similar about previously stated open arithmetic problems. It even seems to be widely

[5]Gödel (1995), p. 307. in his address Gödel suggested that analytic number theory already goes beyond first order number theory. But in later work, Feferman and Takeuti, have independently shown that much of classical analysis can be formalized in a conservative extension of Peano Arithmetic. I am not aware of any investigations of how this applies to analytic number theory.

[6]For some of his early examples see Friedman (1998). His web site https://u.osu.edu/friedman.8/ has many more examples.

accepted, even among logicians, that Gödel undecidability is limited to pathological examples of little general mathematical interest.

It is all too easy to see the unfamiliar as peculiar enough to deserve the label "pathological". Mathematical terminology is littered with unfortunate terminology, residue of what was once though to be pathological: negative numbers, imaginary numbers, irrational numbers, improper integrals. Weierstrass's discovery of an everywhere continuous nowhere differentiable function was greeted with shock. Hermite exclaimed: "Je me détourne avec effroi et horreur de cette plaie lamentable des fonctions continues qui n'ont point de dérivées." (My translation: I turn myself away with fear and horror from this woeful affliction of continuous functions that do not have derivatives.) However, especially with the important role that fractals have assumed such functions no longer induce a shudder of horror. On the other hand seemingly paradoxical consequences of the axiom of choice like that of Banach-Tarski, still make some wince. It should perhaps not be surprising that our intuition should fail us regarding sets too chaotic to be definable in any reasonable language.

A historical example I find particularly revealing was the reaction in the 17th century to Toricelli showing that a certain infinite area when rotated about an axis formed a solid of revolution of finite volume. For the shocked reaction, see Mancosu (1996). This bit of deep pathology, violating all Aristotelian precepts about the necessary separation of the finite from the infinite, is today a homework problem in a freshman calculus course. (The area in question may be taken as formed by the curve $y = 1/x$, the X-axis and the line $x = 1$.)

I venture to predict that the attitude toward Gödel undecidability will eventually undergo a similar change, and that undecidability is destined to play an important role with problems of central mathematical significance.

References

Friedman, H. (1998). Finite functions and the necessary use of large cardinals. *Annals of Mathematics*, *148*, 803893.

Gödel, H. (1995). Some basic theorems on the foundations of mathematics and their implication. In S. Feferman, et al. (Ed.), *Collected works*, (vol. III, pp. 304–323). Oxford.

Mancosu, P. (1996). *Philosophy of mathematics & mathematical practice in the seventeenth century*. Oxford.

Smullyan, R. (1986). *This book needs no title: A budget of living paradoxes*. Paperback: Touchstone Books.

Smullyan, R. (1994). *Diagonalization and self-reference*. Oxford.

Sterne, L. (2003). *The life and opinions of Tristram Shandy, gentleman*, First published 1759–67. Penguin paperback edition.

Chapter 9
Gödel, Lucas, and the Soul-Searching Selfie

Vann McGee

Abstract J. R. Lucas argues against mechanism that an ideal, immortal agent whose mental activities could be mimicked by a Turing machine would be able, absurdly, to prove the Gödel sentence for the set of arithmetical sentences she is able to prove. There are two main objections: "The agent cannot know her own program" and "The agent cannot be sure the things she can prove are consistent." It is argued that accepting the first objection would hand the anti-mechanist a roundabout victory, since for an ordinary finite mechanical system, one can determine what its program is, but that one need not accept the first objection. The second objection can only be thwarted by adopting a conception of "proof" that treats proof as veridical. This reduces Lucas's argument to Montague's theorem on the undefinability of epistemic necessity, which is, it is argued, an obstacle to naturalized epistemology.

9.1 Gödel's "Subjective Mathematics"

The traditional method of acquiring mathematical knowledge has been to derive consequences of axioms regarded as too obvious to require justification. The Gödel incompleteness theorem (1931) presents us with an entirely new method. Start with an evidently true system of arithmetical axioms and go on to conclude the Gödel sentence for that system. We are justified in accepting the Gödel sentence even though it isn't a consequence of the axioms and even though it isn't, or wasn't until it was given a metamathematical gloss, obvious, by anyone's standards. The theorem dramatically reshapes our understanding of the epistemology of mathematics. Do its philosophical implications go even farther, to reshape our understanding of the very nature of the human mind? The possibility was raised by Gödel himself:

> Either mathematics is incompletable in this sense, that its evident axioms can never be comprised in a finite rule, that is to say, the human mind (even within the realm of pure mathematics) infinitely surpasses the powers of any finite machine, or else there exist ab-

V. McGee (✉)
MIT Department of Linguistics and Philosophy, Cambridge, MA, USA
e-mail: vmcgee@mit.edu

M. Fitting and B. Rayman (eds.), *Raymond Smullyan on Self Reference*,
Outstanding Contributions to Logic 14, https://doi.org/10.1007/978-3-319-68732-2_9

> solutely unsolvable diophantine problems of the type specified[1] (where the case that both terms of the disjunction are true is not excluded, so that there are, strictly speaking, three alternatives). (1951, p. 310, emphasis omitted)

The "evident axioms" go beyond the familiar axioms of Peano arithmetic (PA) to include other statements whose truth becomes evident upon mathematical reflection. The deductive closure of the evident axioms forms a system Gödel calls "subjective mathematics," as contrasted with "objective mathematics," which consists in the mathematical propositions that are true. Each of the theorems of subjective mathematics is demonstrable. Whether all these demonstrations can be captured by a single finite system is the question raised by the first disjunct.

To raise the question presupposes that the notion of subjective mathematics, as a system of informal proofs that reaches beyond the confines of any predefined formal system, is sufficiently well-defined that one can reason about it sensibly. In particular, it presumes the ability to distinguish statements that are proved from statements that are merely conjectured and from statements for which one has accumulated a body of evidence that falls short of proof. We shall maintain the presupposition here without examining its merits. It may be that the distinction, while imprecise, is usefully investigated using Van Fraassen (1966) method of supervaluations, a method that has become more-or-less standard among people who wish to accommodate vagueness while upholding the classical rules of deduction. If one wished to employ the method to examine the border of subjective mathematics, a sensible way to proceed would be to begin by assessing how things would play out if the distinction were precise and then ask how vagueness complicates matters. We can think of the discussion here as contributing to the first part of this project. We'll have nothing to say about the second part.

For present purposes, the statements informally proved will be sentences of the language of arithmetic, but there's no prohibition against using ideas from outside arithmetic – from set theory, for instance – in demonstrating that the sentences are true (true, that is, in the standard model).

We'll be discussing "mechanism," the doctrine that the symbol-producing capabilities of the human mind can be simulated by a finite machine. The notion of a finite machine is understood metaphorically, following Gödel (unknown, p. 168), as "a machine with a finite numbers of parts which will do the following thing. If you write down any numbers $n_1,\ldots, n_r$ on a slip of paper and put the slip into the machine and turn the crank, then after a finite number of turns the machine will stop and the value of the function for the argument $n_1,\ldots, n_r$ will be printed on the paper." Whereas we don't have a sharp characterization of what systems count as finite machines, a sharp characterization of the arithmetical functions that are calculable by finite machines is given by the Church-Turing thesis. They are the recursive functions, which are the ones that can be calculated by Turing machines.

[1]The problems are to evaluate statements about the integers of the form "$(\forall y_1)\ldots(\forall y_n)(\exists x_1)\ldots(\exists x_m)p(x_1,\ldots,x_m,y_1,\ldots,y_n)=0$," where p is a polynomial with integer coefficients.

The natural way to think of the mind on a mechanical model is to regard sensory stimuli as inputs and behaviors as outputs. To connect such a model of the mind with Turing machines requires a coding system, since Turing machines have numbers as their inputs and outputs. Developing such a system is a problem we can bypass here. Here (except for §4, where we'll make special arrangements) we are asking about the arithmetical sentences an agent is able informally to prove, working all on her own, without any stimuli. So there are no inputs, and the outputs are arithmetical sentences, which we can identify with their Gödel numbers.

We can't dismiss the possibility that the finite machine that mimics the mind is nondeterministic. A nondeterministic machine calculates f if there is some possible computation on input $< n_1, \ldots, n_r >$ with output $f(n_1, \ldots, n_r)$ and no possible computation with an output different from $f(n_1,\ldots, n_r)$. Nondeterministic machines are modeled by nondeterministic Turing machines, and the functions they calculate are recursive.

Mechanism, as we are understanding it here, requires, given the Church-Turing thesis, that the set of arithmetical sentences a human agent is in a position to accept as established truth be recursively enumerable. Some mechanists go farther, proposing a methodology for psychology that treats the mind as like a computer and its operations as the execution of a program. This methodology isn't required by the meaning of "mechanism" as we're using the word here. It's enough that each mind's outputs be recursively enumerable.

To imagine that the set of arithmetical theses evident to a human mind is closed under logical deduction and that it isn't obviously inconsistent requires a heavy dose of idealization. Human beings are mortal, as are species, and the number of acts, mental or otherwise, we can perform, individually or collectively, is finite. Moreover, we are prone to error. It is unlikely that anyone who has engaged in any significant amount of arithmetical reasoning has been entirely free of computational error, so that the reasoner accepts the results of an incorrect calculation and also conflicting axioms that entail the correct calculation. To get an understanding of "subjective mathematics" that is of any philosophical interest, we need to think counterfactually. What would we accomplish if we enjoyed endless lifespans and boundless mathematical curiosity, our mental faculties were always as acute as humanly possible, and we never committed errors of carelessness or inattention? We ask such questions, not because we care about ideal agents, but because we want to understand the limitations of the methods we and the idealized mathematician both employ.

The mechanistic thesis has it that the human mind, once it's been freed of the limitations of mortality, boredom, fatigue, and careless error, is functionally equivalent to a finite machine that has likewise been freed of mechanical breakdown and power failure. It follows by the Church-Turing thesis that the potential outputs of the mind, in its capacity as seeker after arithmetical truth, form a recursively enumerable set. It's closed under deduction, and so, by Craig (1953), it is the set of consequences of a recursive set of axioms. This recursive axiom system has a Gödel sentence, whose epistemic status is the crux of controversies about the connection between Gödel's theorem and mechanism.

9.2 Lucas's Dialectical Argument

Gödel arrives at a disjunctive conclusion. His sympathies lie with the first, anti-mechanist disjunct,[2] but he doesn't offer a proof of the first disjunct. Lucas (1961, 1968, 1970) proposes to continue where Gödel left off, piecing together Gödel's results so as to arrive at the conclusion Gödel approaches but doesn't reach, that "the human mind ...infinitely surpasses the capacity of any finite machine."

We can characterize the mechanist thesis in terms of a two-move game. First, Player I provides the index[3] e of a recursively enumerable set ω_e. Second, Player II writes down an arithmetical sentence. If the sentence is a member of ω_e our ideal mathematician is unable to recognize as true or it is a nonmember of ω_e the mathematician is able to recognize as true, Player II wins. The mechanist thesis is that Player I has a winning strategy.

Thinking of the mechanist thesis in game-theoretic terms help us understand Lucas's dialectical strategy. His argument is based on the observation that Gödel's proof is constructive. Gödel doesn't merely demonstrate the existence of the undecidable Gödel sentence. He shows how to write it down explicitly. His method requires patience but no ingenuity. You don't have to be an ideal mathematician to carry it out.

The mechanist contends that there is a number e such that ω_e is the set of arithmetical statements the human mind is able informally to prove. The anti-mechanist replies, "Put up or shut up. You claim there is such a number e. Tell me what it is." Unless the mechanist skulks away in embarrassed silence, she'll produce her number e. The anti-mechanist will respond by producing ω_e's Gödel sentence γ and asking whether γ is in ω_e. γ is provably equivalent to the statement that γ is not in ω_e, so that, on the mechanist's account, γ is either false and provable or true and unprovable. The mechanist will surely want to deny that she can prove things that are false, so she'll conclude that γ is true but unprovable. This isn't quite a contradiction, but it's the next worst thing, presenting a statement as proved and in the same breath denying that one is able to prove it. The mechanist position has been shown to be unsustainable, not by *reductio ad absurdum*, but by a closely related form, reduction to pragmatic incoherence. If the agent is sufficiently alert that, when she's proved something, she is able to recognize that she's proved it, the pragmatic infelicity will collapse into outright inconsistency.

A motive for mechanism is the thought that the outputs of our minds are behaviors of a biological organism, and an organism is a finite physical system not fundamentally different from a finite machine. There are substantial differences, of course. The machine was produced by engineers, the organism by natural selection. The two are made of different materials, and the organism is vastly more complex than anything the engineers at Oak Ridge can put together. These differences in architectural detail

---

[2]See Wang (1996, ch. 6).

[3]We follow the notation of Smullyan (1994), in particular, in seeing how to treat "$x \in \omega_y$" as an abbreviation of a Σ_1 arithmetical formula.

are significant, but they are not differences in kind that would make it theoretically impossible to produces a machine functionally equivalent to the organism.

Lucas rejects the picture of the human mind as a component of a physical system that can be understood in purely physicalistic terms, and he thinks the Gödelian argument should induce us to join him in rejecting it. In particular, he advocates a worldview that has room for free will, and a mere machine doesn't have free will.

The physicist Penrose (1989, 1994) took Lucas's line of argument and turned it in a different direction. He has no quarrel with treating the mind as finite physical system, but it's not a system for which familiar mechanical models are adequate. We all know that a mismatch between nature and classical physics is visible at the subatomic level. Penrose thinks that Gödel's theorem shows that the mismatch also appears at the cellular level, and that an adequate science of consciousness will require a marriage of quantum mechanics and neurology.

9.3 The Self-Knowledge Objection

Before we start looking for alternatives to mechanism, we should make sure that the argument against it succeeds. In fact, for all its ingenuity, it doesn't work. The mechanist has been defeated, not by logic, but by peer pressure. She should never have succumbed to the "Put up or shut up" taunt. The mechanist's contention is "There is a Turing machine that can simulate the processes by which the human mind can recognize arithmetical truths." It isn't "There is a Turing machine that can simulate the processes by which the human mind can recognize arithmetical truths, and I can tell you what its program it." An existential claim naturally arouses a desire for further information. Told, "Someone murdered the count," we ask, "Well, who did it?" but the fact that the detective is unable to answer our question is no indication that the answers the detective is able to provide are mistaken. The claim that there are truths of which we will never be aware isn't defeated by the observation that the attempt to provide a witness to the claim would snare the claimant in pragmatic inconsistency.

Someone who contends that there is a Turing program that generates the theorems the human mind is capable of proving is by no means obligated to identify a program that does the job. But suppose that our agent volunteers that information, freely declaring, "There is a program and I can tell you what it is." Then the Lucas argument goes through (although there's a weakness in it that we'll come to presently) and the agent is forced into pragmatic incoherence at best, outright inconsistency at worst. The upshot isn't Lucas's original thesis that the operations of the agent's mind can't be simulated by a machine, but the conditional conclusion that, if there is such a machine, she can't know its program. Paul Benacerraf (1967, p. 30) puts the conclusion dramatically:

> If I am a Turing machine, then I am barred by my very nature form obeying Socrates' profound philosophic injunction: KNOW THYSELF.

9.4 The Soul-Searching Selfie

The anti-mechanist wants to show that the mind is a different sort of thing from familiar machines, not just divergent in details but different in its nature. That aim was the motive for seeking to prove that the mind's activities cannot be simulated by a Turing machine. That effort has faltered. The most the anti-mechanist argument shows is that, if the mind can be simulated by a machine, it isn't a machine that can know its own program. This looks like defeat for the anti-mechanist, but perhaps he can wrest a victory from it. Defeated in the original goal of showing that the mind cannot be mimicked by a machine, he may nonetheless be able to achieve the deeper purpose that motivated the goal, to show that the mind is fundamentally unlike a mechanical system.

For an ordinary mechanical system, like a player piano or jet engine, we can tell how the machine works by removing the cover and examining how it's put together. It would be difficult to do the same thing for a desktop PC, because its parts are so small and delicate, but there's no principled reason why it can't be done. If there's some high-level theoretical reason – a reason based, not in the biological details but in the nature of the mind as a thinking thing – why we can't understand how the mind works the way we understand how a jet engine or a personal computer works, then we will have found a principled difference that sets the human mind apart from ordinary finite physical systems.

Insuperable practical obstacles prevent us from obtaining a detailed mechanical description of the operation of the brain. If we think of the brain as a machine, it's a machine with billions of moving parts, intricately interconnected. To understand its workings, we'd have to look at what's going on inside the cell at the molecular level, and we'd have to do this for billions of cells all at once, without damaging the delicate tissue. Clearly, it can't be done.

It is biologically infeasible to obtain a complete physical description of a living brain. We know this from looking at the brain's physiology. On the assumption that the facts about the mind are determined by the facts about the brain, the Lucas argument as adapted by Benacerraf shows us that is it not merely infeasible but impossible to obtain a full description of the brain, and it shows us this without taking account of anything at all about biology, just on the basis of facts about the human mind that we can discern from the armchair. Chihara's (1972, p. 512) question is apt: "It is natural to suppose that, if I were a Turing machine, it should be at least theoretically possible to discover its program. How can the considerations advanced by Benacerraf put this sort of limit on the discoveries open to the empirical sciences?"

A thorough-going mechanism, one that holds to the spirit as well as the letter of mechanism, requires not only that there be a Turing machine that simulates the operations of the human mind, but also that it be possible, in principle though maybe not in practice, by inspection of the human body, to discover what its program is. We can imagine accomplishing this by an automated process, so that we have a high-tech neuroanalyzer that takes a painless snapshot of a person's brain, analyzes its structure, and gives as output the index of a Turing machine that simulates the person's arithmetical capabilities. If our ideal mathematician turns the machine upon herself, she'll know her own program, so she'll be able to determine her own Gödel sentence, which she'll presumably acknowledge as true. This isn't a problematic outcome, because the state of her brain has been changed by its measurement. We have a mathematician who knows her own program, but not in a threatening way. At a later time she knows what her program was at a earlier time. The threat to mechanism comes if at time t_1 she knows her program at t_1.

The threat can be realized. The mechanist thesis isn't merely that the things a human agent (even an idealized immortal agent) does when she's engaged in solitary contemplation can be mimicked by a Turing machine. The mechanist also contends that the agent's dispositions to respond to environmental stimuli can be described by an algorithm. If we have an effective coding of possible stimuli and potential behaviors, then, according to mechanism, the function that takes the code number of a stimulus to the code number of the agent's response will be recursive.

That's the general situation. Here what we are looking for as a response is not a single act but a whole family of mental acts. We want to determine the set of arithmetical truths the agent will be in a position to prove informally, having received a given stimulus, if she labors forever in her characteristically flawless way, without receiving further stimuli. The mechanist's contention is not only that the set of arithmetically sentences she can informally prove after she's gotten the stimulus will be recursively enumerable, but the process that takes her from the stimulus to the set can be described recursively. The *transition function*, which takes the code number of a stimulus as input and gives an index of her set of personal theorems after she's received the stimulus as output, will be recursive.[4] If tr is the transition function, then, if e is the code of the stimulus, the set of sentences she'll be able to prove informally after getting e will be $\omega_{tr(e)}$. If the thorough-going mechanist is correct in contending that an agent's brain states – and thereby her mental states – are, in principle, subject to detailed physical examination, then it ought to be possible to use the neuroanalyzer to take a snapshot of the agent's brain at time t_0, and thereby discover an index of her transition function at t_0. The device will find a number e such that $tr = q_e$, the eth partial recursive function in a standard enumeration.

To move the adapted version of the Lucas argument forward, we need a gimmick. We need to arrange things so that the agent is able to describe, in detail, the stimulus she's just received. This is something we can't ordinarily do. We can usually give some sort of description of what just happened to us, but we won't be able to characterize our experience in complete, explicit detail. To arrange a situation in which the agent can consciously identify the content of her experience, we resort to an artifice that restricts her range of possible experiences. Suppose that at time t_0 our mathematician is placed in a sensory deprivation chamber, cut off from all sensory stimulation until she gets her next input, which will be an Arabic numeral. To simplify matters, we can take the code number of the experience of being shown the Arabic numeral for n to be n itself.

---

[4] The set G of triples $<x, y, z>$ such that z is the theorem announced by the agent y seconds after receiving x as a stimulus is recursive, so theorem 7.4.1 of Smullyan (1994) gives us our recursive function tr such that, for any x and z, $z \in \omega_{tr(x)}$ iff $\exists y\, G(x, y, z)$.

Before crawling into the sensory deprivation tank, our mathematician sets up the neuroanalyzer with a timer, so that it will take a shapshot of her brain at time t_0 and use the snapshot to calculate a code number of her transition function at time t_0. The neuroanalyzer will then output the answer in the form of an Arabic numeral, presented to the mathematician inside the sensory deprivation tank. Let's say she sees the numeral for e. Then at time t_1, immediately after she's shown the numeral, the mathematician will know that her transition function at t_0 was $q_e(e)$, so that the set of sentences she is in a position to prove at t_1 will be $\omega\, q_e(e)$. She can calculate this, so at t_1 she can determine the program she is using at that very moment and she can produce its Gödel sentence.

That the agent isn't able, just on her own, to determine her own program isn't surprising. Ordinary people have scarcely any aptitude for discerning their own mental states by self-examination. The ideal agent's cognitive skills are our skills flawlessly exercised. It's not to be expected that she will have an ability – detailed introspective discernment of her own mental states – that we so conspicuously lack. It the mechanist is right, the composite system, agent + neuroanalyzer – can be regarded as a two-component mechanical system that does know its own program. This should worry the mechanist.

9.5 The Overconfidence Objection

In his introduction to Gödel (1951), Boolos writes (1995, p. 295), "it is fair to say that" Lucas's argument and its variants "have as yet obtained little credence." This assessment remains accurate today. There are two main lines of attack[5]: "The agent can't know her own program," which is the objection we've been discussing, and "The agent can't be confident that the things she can prove are true," which we'll turn to now. The objection originated in Putnam (1960) response to some remarks of Nagel and Newman (1958) that anticipated Lucas. (We've set aside a third source of complaint, the thought that the notion of informal proof is too mushy to be useful.)

The "She can't know that the things she can prove are true" response is complicated by uncertainty about what is required for an argument to count as a proof. On one reasonable reading of "proof," proof is veridical, since proving a thing is a way of coming to know it, and nothing is known that isn't true. You may sometimes think you've produced a proof when you really haven't, but if you've produced a genuine proof, its conclusion must be true. Fallacious proofs are like counterfeit money: not really money at all. On a veridical conception of "proof," as I understand it, it is built into the meaning of the word "proof" that whatever is proved is true, so that "Whatever is proved is true" is, if I may be allowed to speak this way, analytic. A linguistically competent agent can prove that everything provable is true, just by reflecting on concepts.

[5]In trying to find my way through the literature, I've found Shapiro (1998) especially helpful.

On the concept of proof as veridical, Benacerraf's argument that the agent can't know her own program (assuming she has one) goes through smoothly, applying not only to ideal agents but to ordinary human agents who know enough arithmetic to derive the self-reference lemma. He applies it even to himself: "If I am a Turing machine,..." Of course the set of sentences Benacerraf can prove is recursively enumerable, because it's finite. That's one of the less disturbing consequences of human mortality. If we accept Smullyan's assessment (1992, p. 118) regarding mortality that "little things like that don't bother us in mathematics," we can ask about the set of sentences he could prove if he had the opportunity to go about his work forever. We find that the considerations that make mechanism attractive give us no reason to imagine that the set of sentences he can prove will be recursively enumerable. If mechanism is correct, the cognitive activities of a human agent, even a hypothetical immortal one, can be simulated by a Turing machine, which implies that the set of sentences that the agent places in the tray marked "Proved" is recursively enumerable. However, if the agent is fallible, it's likely that some of the sentences the agent places in the tray will be untrue, and so not proved. Mechanism tells us that the agent's outputs, the things that eventually appear in the tray marked "Proved," will form a recursively enumerable set. However, the sentences that are actually proved will form a subset of the items that appear in the tray, and we've no reason to suppose that the subset is recursively enumerable. Benacerraf showed that, if there is a Turing machine that generates the sentences an agent is able to prove, the agent can't know its program. The conditional conclusion loses much of its interest when we realize that mechanism gives us no good reason to accept the antecedent.

There is a more liberal usage of the word "proof," according to which a fallacious proof is like a rotten egg: rotten but still an egg. Let's not get bogged down in a verbal dispute. We can say this: from the alleged fact that the agent's mathematical activities can be simulated by a Turing machine, we can conclude that the set of sentences the agent accepts as proved is recursively enumerable, but we cannot automatically assume that it is a subset of Gödel's objective mathematics. This is crucial. The Gödel sentence for a recursively axiomatized system that includes Robinson's Q[6] is provably equivalent to the statement that the system is consistent, so if the sentences that appear in the "Proved" box are inconsistent, the agent's Gödel sentence will be false.

Human beings are prone to error. It is a moral certainty that, for any human being who has worked with numbers enough to have a significant body of arithmetical beliefs, the set of these beliefs is inconsistent. Human mathematicians are highly fallible, but are the methods of mathematics fallible? The alternative is to suppose that, whenever a mathematician reaches a false conclusion by attempting to apply standard mathematical methods, it's the fault of the mathematician and not the methods.

Within the empirical sciences, one can reason as well as humanly possible, and still reach an incorrect conclusion, because one has been presented with misleading evidence. An epistemically blameless scientist can come to a false conclusion. It happens all the time. Does the same thing happen within mathematics, or is it the

[6] From Tarski et al. (1953, p. 51).

case that, whenever a mathematician arrives at a false conclusion, it's because he's committed an identifiable blunder? The standard pattern is that, when mathematicians observe a pattern, they'll conjecture that the pattern holds generally, but they won't elevate the conjecture to the status of an established fact until they have a proof. Empirical scientists can't afford to be so demanding.

Our idealized mathematician doesn't make careless errors. Before she presents a statement as proved, she goes over it so carefully that, had there been an identifiable error, she would have noticed it. If it's the case that, in mathematics, unlike the empirical sciences, one never reaches a false conclusion blamelessly, the arithmetical statements our idealized mathematician accepts as proved will, in fact, be true.

Let us suppose that our idealized mathematician has this much self-knowledge: she knows that, in her mathematical reasoning, she is never careless or hasty. She methodically checks her work to make sure that she hasn't committed any calculating errors or deductive mishaps, and in developing a proof, she always takes the time to fill in all the steps. Suppose moreover that she subscribes to the doctrine that no one ever arrives at a false mathematical conclusion blamelessly. Whenever there is a false conclusion, there is an error that could and should have been avoided. In this situation, the agent can reach the extraordinary, but in the circumstances not unreasonable, conclusion that all the arithmetical sentences she comes to accept as proven will, in fact, be true. For any ordinary human, such cockiness would be ludicrous, but our idealized agent's self-assurance is a product, not of vainglory, but of candid assessment.

The strategy of assuring that an agent's purported proofs are sound by adopting especially rigorous standards of proof is in danger of defeating itself. To carry out the Gödel construction, it is not enough that the sentences the agent presents as proved – the agent's outputs – are all true. We need, in addition, that the agent is able to present as proved the judgment that the sentences the agent presents as proved are true (or at least consistent; more on this below). By raising her standard for what she's willing to count as a proof, she makes it more likely that the things she can prove are all true, but at the same time, she makes it more difficult to count this assessment, "Everything I am able to prove is true," as proved. The sweet spot – rigorous enough that she can be confident that everything she can prove is true, but relaxed enough to count her grounds for this confidence as constituting a proof – is elusive.

The statement that everything in ω_e is true is naturally expressed in the metalanguage, as "$\forall x(x \in \omega_e \supset \mathrm{Tr}(x))$," but we can achieve the same effect without leaving the object language by taking the reflection conditionals ($\ulcorner\varphi\urcorner \in \omega_e \supset \varphi$) as axiomatic. More precisely, given a background arithmetical theory Γ,[7] the arithmetical consequences of $\Gamma \cup$ {(T)-sentences ($\mathrm{Tr}(\ulcorner\varphi\urcorner) \equiv \varphi$), for φ arithmetical} $\cup$ {$\forall x(x \in \omega_e \supset \mathrm{Tr}(x))$} will be the arithmetical consequences of $\Gamma\cup$ {reflection conditionals ($\ulcorner\varphi\urcorner \in \omega_e \supset \varphi$), for φ arithmetical}.

Suppose we have an agent who is functionally equivalent to a Turing machine and who is both self-aware and self-confident. Her self-awareness means that she

[7] We assume that Γ contains enough syntactic theory to ensure that distinct sentences have distinct Gödel codes. Halbach (2001) stresses the need for this assumption.

can identify a recursively enumerable set ω_e as the code numbers of sentences she is in a position to accept as proved. Her self-confidence means that she accepts it as proved that the elements of ω_e are true. This means that she will accept the following sentences:

$(\ulcorner\varphi\urcorner \in \omega_e \supset \varphi)$, for φ an arithmetical sentence.

$\ulcorner(\ulcorner\varphi\urcorner \in \omega_e \supset \varphi)\urcorner \in \omega_e$, for φ an arithmetical sentence.

$\ulcorner\theta\urcorner \in \omega_e$, for θ an axiom of logic[8] or an axiom of Q.

$(\ulcorner(\varphi \supset \psi)\urcorner \in \omega_e \supset (\ulcorner\varphi\urcorner \in \omega_e \supset \ulcorner\psi\urcorner \in \omega_e))$, for φ and ψ arithmetical sentences.

Her confidence is misplaced. The set is inconsistent with Q. For a proof, take γ to be the Gödel sentence for ω_e, and note that the following sentences are all consequence of the set together with Q:

$\ulcorner(\ulcorner\gamma\urcorner \in \omega_e \supset \gamma)\urcorner \in \omega_e$	[reflection conditionals are in ω_e]
$\ulcorner(\ulcorner\gamma\urcorner \notin \omega_e \supset \gamma)\urcorner \in \omega_e$	[consequences of Q are in ω_e]
$\ulcorner\gamma\urcorner \in \omega_e$	[ω_e is closed under consequence]
$(\ulcorner\gamma\urcorner \in \omega_e \supset \gamma)$	[a reflection conditional]
$(\ulcorner\gamma\urcorner \in \omega_e \supset \sim\gamma)$	[a consequence of Q]
$(\gamma \wedge \sim\gamma)$	[tautological consequence of the previous three lines]

The conclusion we reach is that a Turing machine that knows its own program can't be fully confident of its own reliability. This is Benacerraf's conclusion, with his assumption that the machine's outputs are known to always be true (which he makes mainly as a concession to Lucas) made explicit.

9.6 Consistency and Reflection

The assumption, in the argument just given, that the agent accepts all the reflection conditionals as proven was substantially more than was really needed. If we look back at Gödel's Gödel (1931) paper, we see that his proof doesn't require the hypothesis that the theory under discussion be true, only that it be consistent. In other words, the only reflection conditional needed is $(\ulcorner 0 \neq 0\urcorner \in \omega_e \supset 0 \neq 0)$. If our agent is self-assured enough to be confident that the sentences she can prove are consistent, the following sentences will be among the statements she embraces:

[8]For simplicity, we take the axioms of logic to be the system of Quine (1951, §15), whose sole rule of inference is *modus ponens*.

$\ulcorner 0 \neq 0 \urcorner \notin \omega_e$.

$\ulcorner\ulcorner 0 \neq 0 \urcorner \notin \omega_e \urcorner \in \omega_e$.

$\ulcorner \theta \urcorner \in \omega_e$, for θ an axiom of logic or an axiom of PA.

$(\ulcorner(\varphi \supset \psi)\urcorner \in \omega_e \supset (\ulcorner \varphi \urcorner \in \omega_e \supset \ulcorner \psi \urcorner \in \omega_e))$, for φ and ψ arithmetical sentences.

$\ulcorner(\ulcorner(\varphi \supset \psi)\urcorner \in \omega_e \supset (\ulcorner \varphi \urcorner \in \omega_e \supset \ulcorner \psi \urcorner \in \omega_e))\urcorner \in \omega_e$, for φ and ψ arithmetical sentences.

Again, her confidence is misplaced. The set is inconsistent with PA. This result is just a restatement of the second incompleteness theorem. The key observation underlying the proof is that, when a number a is an element of a recursively enumerable set ω_d, then the computation that put a into ω_d can be described in Q, so that $\ulcorner \underline{a} \in \omega_d \urcorner$ is a consequence of Q and hence an element of ω_e. This observation can be formalized in PA, so that the conditional $(\underline{a} \in \omega_d \supset \ulcorner \underline{a} \in \omega_d \urcorner \in \omega_e)$ is a consequence of PA and hence an element of ω_e. In particular, $\ulcorner(\ulcorner \gamma \urcorner \in \omega_e \supset \ulcorner\ulcorner \gamma \urcorner \in \omega_e \urcorner \in \omega_e)\urcorner \in \omega_e$.

The assumption that the sentences that are provable are true is extravagant, but it's not unmotivated. We've found two reasons for thinking that the set of provable sentences is consistent: general considerations about the meaning of the word "prove," which lead us to think that proof is veridical, and particular considerations about the intellectual virtues of our agent, which lead us to think she is reliable. Both considerations support the stronger hypothesis that the provable sentences are all true.

Our agent utilizes the standard methods of mathematics, and she employs them with patience, rigor, and care. If she knows this about herself, she'll have reason to think that everything she proves mathematically is true. If she has identified ω_e as the set of sentences she is able to prove mathematically, then she'll have reason to accept the reflection conditionals for ω_e. There is a crack here where doubt can seep in. If she uses a mechanical device for determining what her program is, how can she be sure the equipment hasn't malfunctioned? If she determines her program by introspection, how can she be sure that her introspective faculties are accurate when the introspective abilities of ordinary mortals are so impaired? Hasn't she read Wittgenstein?

Even if she is somehow completely confident that ω_e is her program, the reflection conditionals for ω_e won't be things she's proved by the standard methods of mathematics, so if ω_e consists of the sentences she is able to prove by standard mathematical methods, the reflection conditionals won't be elements of ω_e.[9] It may be that she's confident enough of her own accuracy in applying the standard methods of mathematics and of her correctness in identifying ω_e as contained within the set of things she is able to prove by standard methods that she is willing to regard the reflection conditionals for ω_e as informally proved. If we let $\omega_{\text{Ref}(e)}$ be the set of consequences of $\omega_e \cup \{$reflection conditionals for $\omega_e\}$,[10] she'll regard the members of $\omega_{\text{Ref}(e)}$ as proved. She'll accept the legitimacy of the rule "Accept any arithmetical

[9]Gaifman (2000) emphasizes this point.

[10]The function Ref is given by Theorem 7.4.1 of Smullyan (1994).

statement that can be derived by the standard methods of mathematics supplemented by the reflection conditionals for the set of sentences I can prove by the standard methods of mathematics," so she'll want to accept the reflection conditionals for $\omega_{\text{Ref}(e)}$, even though they're not obtained by the standard methods of mathematics supplemented by the reflection conditionals for the set of sentences she can prove by the standard methods of mathematics. To capture them, she'll have to go to the still stronger theory $\omega_{\text{Ref}(\text{Ref}(e))}$.

The process that takes us from ω_d to $\omega_{\text{Ref}(d)}$ isn't conservative. Unless ω_d contains all sentences, $\omega_{\text{Ref}(d)}$ will be strictly larger than ω_d. The operation is soundness-preserving, that is, whenever ω_d generates only sentences that are true in the standard model, $\omega_{\text{Ref}(d)}$ does the same. If you've identified ω_e as the set of sentences you could prove at t_0, you can get to $\omega_{\text{Ref}(e)}$ by adopting the new rule, "You may assert reflection conditionals for the set of sentences you were able to prove at t_0." The process can be extended into the transfinite. Feferman (1962) investigation of iterated reflection conditionals, which follows in the footsteps of Turing (1939) investigation of iterated consistency statements, is one of the high-water marks of modern logic.

Starting with a recursively enumerable set of sentences that can be proved by standard mathematical methods and repeatedly applying the Ref operation will never get us a fixed point, an index k for which $\omega_k = \omega_{\text{Ref}(k)}$. A fixed point is what you'll get if you have the rule, "You may assert reflection conditionals for the system of rules you are employing right now (a system that includes this very rule)." Even though the rule refers to itself, we can get a fixed point by a standard self-referential technique, namely, Kleene's recursion theorem.[11] However, the only fixed point we get is $\omega_k = \{\text{sentences}\}$, which corresponds to the rule, "Assert anything you like."

The promise that an ideal agent could, by reflecting on her own intellectual virtues, devise a theory that affirms its own truth, or at least its own consistency, is illusory. Reflection on the soundness of the methods that gave her the axiom system ω_e may give her powerful reasons to believe the reflection conditionals for ω_e. But having powerful reasons to believe something isn't the same as having a proof.[12] To count the reflection axioms for ω_e as proved, she'll need to add them as further axioms, and doing so will again leave her with a system unable to prove its own consistency.

If mechanism is correct, the set of sentences an agent is in a position to count as proved – the sentences she can put into the tray marked "Proved" – will be a recursively enumerable set. The fully committed mechanist will admit that, in certain exotic circumstances, it will be possible for the agent to specify the recursively enumerable set. For an ideally meticulous agent, the sentences she puts into her "Proved" box will be the sentences she has, in fact, proved, and so, assuming that whatever is proved is thereby known, they will be sentences she knows to be true. Moreover, if the agent is aware of the care and rigor with which she undertakes her own work, she'll have convincing reason to believe that the sentences she puts into her "Proved" tray are all proved, and so all true. This convincing reason falls short of a proof, by her own exacting standards of proof. Much as she believes that the

[11] Smullyan (1994, p. 260f).

[12] This observation was emphasized by Shapiro (1998).

sentences that appear in her "Proved" tray are all true, this isn't something she's proved, and as long as she refrains from putting claims of the verity of the contents of her "Proved" tray into her "Proved" tray, she'll avoid contradicting herself, and the attack on mechanism will be thwarted. Fending off the attack does not, of course, establish the correctness of mechanism.

9.7 The Inexpressibilty of Provability

Gödel's proof that a consistent theory that includes PA doesn't prove its own consistency relied heavily on the supposition that the set of provable sentences was a recursively enumerable set identified by a Σ_1 formula.[13] This is what we relied on in establishing that, if $\ulcorner\gamma\urcorner \in \omega_e$, then it is provable that $\ulcorner\gamma\urcorner \in \omega_e$. However, if we look at the simpler proof that started with the reflection conditionals rather than the consistency statement, we see that assumption of Σ_1-definability was never used. Take any arithmetical formula $\chi(x)$ and replace "$x \in \omega_e$" with "$\chi(x)$" Constructing the Gödel sentence γ for χ, for which Q proves $(\gamma \equiv\sim \chi(\ulcorner\gamma\urcorner))$, the argument shows that the following sentences are inconsistent with Q:

$$(\chi(\ulcorner\varphi\urcorner) \supset \varphi), \text{ for } \varphi \text{ an arithmetical sentence.}$$

$$\chi(\ulcorner(\chi(\ulcorner\varphi\urcorner) \supset \varphi)\urcorner), \text{ for } \varphi \text{ an arithmetical sentence.}$$

$$\chi(\ulcorner\theta\urcorner), \text{ for } \theta \text{ an axiom of logic or an axiom of Q}$$

$$(\chi(\ulcorner(\varphi \supset \psi)\urcorner) \supset (\chi(\ulcorner\varphi\urcorner) \supset \chi(\ulcorner\psi\urcorner))), \text{ for any } \varphi \text{ and } \psi.$$

This theorem is one of Montague (1963) results on the syntactical treatment of modality.[14] On a way of using the word "prove" so that proof implies truth, provability is a species of epistemic necessity.

If mechanism is correct, the set of sentences an agent is in a position to accept as proved will be the extension of a Σ_1 formula. For an ideally meticulous agent, there are no discrepancies between the sentences the agent is in a position to accept as proved and those she is, in fact, in a position to (veridically) prove. So, according to the mechanist, the set of sentences the ideal agent is capable of proving is the extension of a Σ_1 formula. However, the set of provable sentences isn't *provably* the extension of any Σ_1 formula, or of any arithmetical formula.

It's starting to look as if the issue of mechanism was something of a red herring. On a conception of "proof" that allows proof without truth, the attempt to derive, on

---

[13]In presenting the second incompleteness theorem above, we gave five conditions ("$\ulcorner 0 \neq 0\urcorner) \notin \omega_e$," etc.) that are incompatible with PA. If, in the five conditions, we replace the Σ_1 formula "$x \in \omega_e$" with the Σ_2 formula "$\exists n \exists s$ (s is a derivation of x from axioms of PA with code numbers $<n$, and the set of axioms of PA with code numbers $<n$ is consistent)," which is an alternative way of identifying the set of consequences of PA, the substitution transforms the five conditions into theorems of PA.

[14]See also Montague and Kaplan (1960).

the assumption that mechanism is correct, a contradiction from the hypothesis of an ideal agent who is able fully to specify the sentences she is able to prove founders on the "She can't prove that the sentences she can prove are consistent" objection. On a concept of proof as veridical, the efforts succeeds, but the hypothesis of mechanism no longer plays a role.

An unnamed friend of Benacerraf, quoted in his (1967, p. 30), responded to Benacerraf's version of the Lucas argument by declaring that psychology as we know it is impossible. The reaction seems overheated. We ask psychologists to describe the mind and its capabilities, but we don't demand that they describe it in complete detail. Elsewhere in the sciences, we are happy if the scientists can give an account of the object of their inquiry that is accurate and informative; we don't demand that it also be exhaustive. There is no reason to place more demanding requirements on psychologists.

Montague's paradox is more troubling, for it raises the possibility that our psychological theories aren't merely doomed to be incomplete, they are doomed to be incoherent. An ambitious program for psychology has it take on many of the tasks traditionally assigned to epistemology, so that it aims not only to describe the mechanisms by which agents form beliefs, but to tell us when and how those mechanisms succeed in giving the agents knowledge. This is Quine (1969) program of naturalized epistemology. It aims to distinguish, among the modes of demonstration that lead agents to accept a thesis as proved, those modes that succeed in establishing the truth of the conclusion. It is this ambitious program for psychology, with its veridical understanding of "proof," that is threatened by Montague's result.

On an understanding of proof as veridical, we would expect an account of what a person is ideally capable of proving to include the following theses, assuming it can be formalized within a first-order language $\mathscr{L}$ that incorporates the language of arithmetic as a way of encoding syntax:

$(\ulcorner\varphi\urcorner \text{ is provable } \supset \varphi)$, for φ a sentence of $\mathscr{L}$.
$\ulcorner(\ulcorner\varphi\urcorner \text{ is provable } \supset \varphi)\urcorner$ is provable, for φ a sentence of $\mathscr{L}$.
$\ulcorner\theta\urcorner$ is provable, for θ an axiom of logic or of Q.
$(\ulcorner(\varphi \supset \psi)\urcorner \text{ is provable } \supset (\ulcorner\varphi\urcorner \text{ is provable } \supset \ulcorner\psi\urcorner \text{ is provable}))$, for any φ and ψ in $\mathscr{L}$.

Our expectations are thwarted. The theses are inconsistent with Q.

Tarski (1935) showed that the (T)-sentences $(\text{Tr}(\ulcorner\varphi\urcorner) \equiv \varphi)$ are inconsistent with the laws of syntax. The reflection conditionals are the left-to-right direction of the (T)-sentences, with "Tr" replaced by "provable." The supposition that the reflection conditionals are true and provable gives provability enough of the formal properties of truth to allow the emergence of semantic paradox.

If there can be said to be an orthodox response to the liar paradox, it is Tarski's: the truth theory for a language should be developed, not in the language itself, but in a richer metalanguage. We can apply the same mandate to provability, although doing so dims our hope for a comprehensive psychology, since the demonstrative methods we employ in giving proofs about proofs will lie outside the field of vision of our theory of proofs.

We can follow Tarski farther. He proposed replacing "Tr" with a hierarchy "Tr_0," "Tr_1," "Tr_2,"...of truth predicates, understood so that the extension of "Tr_n" consists of those sentences φ that don't contain any of the predicates "Tr_k," with $k \geq n$, and that satisfy the biconditional $(\mathrm{Tr}_n(\ulcorner\varphi\urcorner) \equiv \varphi)$. We can do the same thing with a hierarchy of provability predicates "$\mathrm{provable}_0$," "$\mathrm{provable}_1$," "$\mathrm{provable}_2$," ..., where the extension of "$\mathrm{provable}_n$" excludes sentences that contain "$\mathrm{provable}_k$" with $k \geq n$. Tarski's work shows that the resulting theory avoids paradox, by producing a model which describes the special case of provability by an omniscient agent.

This consistency result is reassuring, but I can't shake the feeling that Benacerraf's friend was onto something. If our understanding of the human capacity for demonstrative reasoning is unavoidably fragmented and partial, the prospects for a comprehensive science of human cognition are rather limited.

Montague recommended a different approach from Tarski's, expressing provability not with a predicate or a sequence of predicates, but with an operator "$\Box$." We get the simplest application of his method by going back to the language of arithmetic and forming a Gödel-style provability predicate "Bew" for a consistent, recursively axiomatized theory that includes PA. We can introduce the epistemic modal operator into the language by stipulating that $\Box\varphi$ is to be a sentence whenever φ is a sentence, and once we've introduced the operator, we can eliminate it, working from the inside out, by applying the stipulation $\Box\varphi =_{\mathrm{Def}} (\mathrm{Bew}(\ulcorner\varphi\urcorner) \wedge \varphi)$. The valid modal formulas will be the theorems of S4, supplemented with Grzegorczyk's axiom:

$$(\Box(\Box(\varphi \supset \Box\varphi) \supset \varphi) \supset \varphi).$$

See Boolos (1993, ch. 12). If γ is the Gödel sentence for "Bew," for which we have $(\gamma \equiv\sim \mathrm{Bew}(\ulcorner\gamma\urcorner))$, we find that the biconditional $(\gamma \equiv\sim \Box\gamma)$ is true but unprovable, and that there is no sentence λ for which $(\lambda \equiv\sim \Box\lambda)$ is a theorem. So paradox is averted.

There are numerous other possible responses, closely parallel to responses to the liar paradox. Many of them require relinquishing classical logic. I expect that a fully satisfactory response to the paradox of veridical provability will await a satisfactory response to the liar paradox, and that it will be a long wait.

References

Benacerraf, P. (1967). God, the devil, and Gödel. *The Monist*, *51*, 9–32.

Boolos, G. S. (1993). *The unprovability of consistency*. Cambridge: Cambridge University Press.

Boolos, G. S. (1995). Introduction to Gödel (1951) in Gödel, *Collected works*, vol. III (pp. 290–304). Oxford: Oxford University Press.

Chihara, C. (1972). On alleged refutations of mechanism using Gödel's incompleteness results. *Journal of Philosophy*, *69*, 507–536.

Craig, W. (1953). On axiomatizability within a system. *Journal of Symbolic Logic*, *18*, 30–32.

Feferman, S. (1962). Transfinite recursive progressions of axiomatic theories. *Journal of Symbolic Logic*, *27*, 259–316.

Gaifman, H. (2000). What Gödel's incompleteness result does and does not show. *Journal of Philosophy*, *97*, 462–471.
Gödel, K. (1931). Über formal unentscheidbare Sätze der Principia mathematica und verwandter Systeme I. *Monatshefte für Mathematik und Physik*, *38*, 173–198. Trans (English). J. van Heijenoort in van Heijenoort (Ed.), *From Frege to Gödel*, (pp.596–626) (Cambridge, MA: Harvard University Press, 1967), and Gödel (Ed.), *Collected works*, vol. I, (pp. 144–195) (Oxford: Oxford University Press, 1986).
Gödel, K. (unknown). Undecidable diophantine polynomials. In Gödel, *Collected works*, vol. III (pp. 164–174). Oxford: Oxford University Press, 1995.
Gödel, K. (1951). Some basic theorems on the foundations of mathematics and their implications. In Gödel (Ed.), *Collected works*, vol. III (pp. 304–323). Oxfod: Oxford University Press, 1995.
Halbach, V. (2001). How innocent is deflationism? *Synthese*, *126*, 167–194.
Lucas, J. R. (1961). Minds, machines, and Gödel. *Philosophy*, *36*, 112–127.
Lucas, J. R. (1968). Satan stultified. *The Monist*, *52*, 145–158.
Lucas, J. R. (1970). *Freedom of the will*. Oxford: Oxford University Press.
Montague, R. (1963). Syntactical treatments of modality, with corollaries on reflexion principles and finite axiomatizability. *Acta Philosophica Fennica*, *16*, 153–167. Reprinted in Montague. *Formal philosophy* (pp.286–302). New Haven: Yale University Press, 1974.
Montague, R., & Kaplan, D. (1960). A paradox regained. *Notre Dame Journal of Formal Logic*, *1*, 79–90. Reprinted in Montague, Formal philosophy (pp. 271–285). New Haven: Yale University Press, 1974.
Nagel, E., & Newman, J. R. (1958). *Gödel's proof*. New York: New York University Press.
Penrose, R. (1989). *The emperor's new mind*. Oxford: Oxford University Press.
Penrose, R. (1994). *Shadows of the mind*. Oxford: Oxford University Press.
Putnam, H. (1960). Minds and machines. In S. Hook (Ed.), *Dimensions of mind* (pp. 148–179). New York: New York University Press. Reprinted in Putnam, *Mind, Language and Reality, Philosophical Papers*, vol. 2 (pp.362–385). Cambridge: Cambridge University Press, 1975.
Quine, W. V. (1951). *Mathematical logic*, revised ed. New York: Harper & Row.
Quine, W. V. (1969). Epistemology naturalized. In *Ontological relativity and other essays* (pp. 69–90). New York: Columbia University Press.
Shapiro, S. (1998). Incompleteness, mechanism, and optimism. *Bulletin of Symbolic Logic*, *4*, 273–302.
Smullyan, R. M. (1992). *Gödel's incompleteness theorems*. Oxford: Oxford University Press.
Smullyan, R. M. (1994). *Diagonalization and self-reference*. Oxford: Oxford University Press.
Tarski, A. (1935). Der Wahrheitsbegriff in den formalisierten Sprachen. *Studia Philosophica*, *1*, 261–405. Trans (English) J. H. Woodger in Tarski (Ed.), *Logic, semantics, metamathematics*, 2nd ed (pp. 152–178). Indianapolis: Hackett, 1983.
Tarski, A., Mostowski, A., & Robinson, R. M. (1953). *Undecidable theories*. Amsterdam: North-Holland.
Turing, A. M. (1939). Systems of logic based on ordinals. *Proceedings of the London Mathematical Society*, ser. 2, *45*, 161–228, Reprinted in M. Davis (Ed.), *The Undecidable* (pp. 115–222). San Diego: Raven Press, 1965.
Van Fraassen, B. C. (1966). Singular terms, truth value gaps, and free logic. *Journal of Philosophy*, *63*, 464–495.
Wang, H. (1996). *A logical journey: From Gödel to philosophy*. Cambridge: MIT Press.

Chapter 10
An Island Tale for Young Anthropologists

Andrew G. Buchanan and John H. Conway

Abstract In which the fictional Professor Kit Sune explores the Archipelago of Ambiguity, following in the footsteps of an earlier explorer, Raymond Smullyan. Here, she reviews the great progress made in Knight-Knave studies since Smullyan's seminal work. In the process, she discovers an eight word question she can repeat three times to identify three natives' tribal affiliations. She also gives an answer to a knotty problem posed by Simon Norton. After pausing to gasp in horror at the looming "Knownot Apocalypse", she finishes by posing 14 open Knight-Knave questions.

10.1 In Which We Meet Professor Sune

Let us introduce you to the esteemed anthropologist, Professor Kit Sune,[1] a specialist in the mythology of tricksters. She is an expert in heroes (like Br'er Rabbit and Odysseus) and in rogues (like Loki and Wile E. Coyote). Truth and lies fascinate her equally, so where better for her to conduct research than the fabled Archipelago of Ambiguity?

Here, she follows in the footsteps of an earlier explorer, the logician Raymond Smullyan. Curiously, he has chosen not to report his fieldwork in mainstream anthropological journals, instead recounting his adventures in fictionalized form across a number of very popular puzzle books (Smullyan 1978, 1982a, b, 1985, 1987, 1992, 1997, 2007, 2008, 2013).[2]

A. G. Buchanan (✉)
Anselan Ltd, Sha Tin, Hong Kong, China
e-mail: andrew@anselan.com

J. H. Conway
Department of Mathematics, Princeton University, Fine Hall, Princeton, USA

[1] *Kit Sune* is the generic Japanese word for fox, but has a more specific meaning in folklore: a wise, magical, vulpine trickster, generally benevolent, with multiple tails, and often shape-shifting into human form.

[2] On p. 147 of "*What is the Name of this Book?*", Smullyan, having deliberately backed himself into a narrative corner, states that an island he had been discussing as fact is, in fact, fictional. However we are not to be fooled by that. In the preface to the book, he hopes that it will enable his wife to decide whether she is married to a Knight or a Knave. From p. 147, we conclude that he is neither. Hence the island may exist.

M. Fitting and B. Rayman (eds.), *Raymond Smullyan on Self Reference*,
Outstanding Contributions to Logic 14, https://doi.org/10.1007/978-3-319-68732-2_10

A visitor to an island in the Archipelago is conventionally welcomed at the jetty by a friendly Reception Committee comprising exactly one representative of each resident tribe. Two tribes are widespread in the Archipelago: one whose members always tell the truth, and another whose members always lie. (Tell us if you've heard this one before!) (Gardner 2013) Smullyan named these tribes *Knights* and *Knaves* respectively in his books, and so shall we call them here. Recently however, the little schooner in which Professor Sune is voyaging has reached a region also populated by a third tribe: the *Knownots*, who sometimes tell the truth and sometimes lie, whichever suits them at the time. (They are like Smullyan's *Spies* or *Normals*.)

On arrival, time-honoured protocol demands that the anthropologist, knowing which tribes dwell on the island, questions the Reception Committee to deduce the tribe of each member. Questions should be binary (i.e. with only two possible answers). It is polite to pose the minimum number of questions, because everyone is eager to commence the merry Welcoming Party which will follow. It is respectful, however, to determine the tribal affiliations fully through logical inference, rather than guesswork.

10.2 In Which Professor Sune Shows Her Resourcefulness in the Face of Adversity

Professor Sune's original plan was to ask twice a Key Question (KQ): "Would you please point to which of your committee colleagues is less likely to tell the truth?",[3] addressing the second question to the person indicated by the first. Finally, to the person now indicated, she would pose a Last Question (LQ): "Are you a tree-frog?"[4] The point of KQ is that the person pointed to cannot be a Knownot, since the Knight points to the Knave, the Knave points to the Knight and the Knownot points to either the Knave or the Knight. So after KQ has been asked twice, two distinct people have been indicated neither of whom can be the Knownot, so by elimination the third is the Knownot. Then LQ is addressed to either the Knight or the Knave, to identify which is which.

As the schooner approached the jetty, Sune's only remaining task was to use her satellite phone to translate KQ and LQ into the island's dialect. (The English words "yes" and "no" are universal in the archipelago, as is the concept of pointing.) However, she suddenly realized that her phone battery level was extremely low. She assessed that she would only have time to translate one question. Her only chance was therefore to find a single terse binary question that she could repeatedly pose.

At first, it seemed impossible, because surely each repetition would have to be posed to a different individual. The Knownot's answer would contribute no informa-

[3] Grammatically an indirect question, but a binary question nonetheless.

[4] LQ can obviously be replaced by any question to which we know the answer, e.g. "Does $1+1=2$?". The tree-frog question is taken from a wonderful scene in Werner Herzog's movie "*The Enigma of Kaspar Hauser*" (Herzog 1974).

tion, and there are only $2 \times 2 = 4$ ways that the other two could respond, insufficient to distinguish the six possible assignments of tribe to the Reception Committee members. You might like to solve this challenge before reading on.

Sune managed to find a question with only eight words and some pointing: "Does *this* person lie more than *that* person?" She made this question equivalent to KQ the first two times by pointing to the two committee members other than the questionee. However, on the third asking, she made it equivalent to LQ by pointing to the same person twice!

"But isn't that cheating?" you may think. Can a single question have "variables" in it? Smullyan addressed this on p. 29 of "*To Mock a Mockingbird*". He posits two brothers, one honest and sane and the other dishonest but deluded. The sane see the world truly, while the deluded are incorrect in every belief. Hence for example: "Does $1+1=2$?" gets the answer "yes" from both brothers. But Smullyan sneakily observes that there *is* a way to distinguish them. If you ask "are you honest?", the honest brother will say "yes", while his sibling, via a treble negative, will say "no". They give different answers because the questions are not identical. The word "you" translates to a different brother in each case. As Smullyan wrote:

> And so you are not really asking the same question, even though the sequence of words is the same. The technical adjective for words like "you", "I", "this", "that" and "now" is **indexical**. Their denotations are not absolute, but depend upon their contexts.

Our interpretation of "sameness" here is looser but still natural: *questions are the same if only one translation was required on Sune's mobile phone.*

To avoid pointing, we could instead decorate members of the committee, with e.g. hat, sunglasses and floral garland, and move these objects around between questions. "O person in hat: does the person in sunglasses lie more than the person wearing the garland?" We think pointing is simpler.[5]

Anyway, enough hair-splitting: let's pick up the story again.

10.3 In Which Professor Sune Learns Something Surprising About the Island's Recent History

Following the Professor's success, tea was served. With her trained anthropologist's observation skills, Professor Sune noted that participants were not asked whether they desired milk, lemon or sugar, but were simply allowed to serve themselves.

One of the committee interrupted her study: "It was nice to see an old school, hand-crafted question strategy."

---

[5]For those who still think it's rude to point, there is another very different approach. For example: "Is it true that either I have asked this question of you before or that the taller other member of the Reception Committee lies more often than the shorter?" (Norton 2015). This implicitly uses a "state" indexical to record the history. Using this general strategy one could embed within a single (long) question a complete interrogation of 1000 islanders.

"What do you mean?" asked the Professor, checking that it was indeed the Knight who had just spoken.

(It was actually quite easy to guess what tribe someone came from, she realized. The Knights looked comfortable and at ease, after a lifetime of being taken at their word. The Knownots had the hangdog look of someone who was never listened to. And the Knaves often spoke woodenly, simply reversing statements that were true. In her inferences, she always ignored such psychological cues, however.)

"Well..." began the Knownot.

The Knave interrupted: "We can't ignore the Knownot. There haven't been large teams of anthropologists visiting us recently. They have not employed what they termed a 'scalable big data paradigm'".

"And so you all can speak fluent English now?" Sune asked, sad that her time fiddling with the mobile phone had apparently been for nothing.

"Yes!" "No!" "Yes!" replied the Committee.

"Come and see the other side of the island," said the Knight, "as soon as we've finished our tea."

10.4 In Which Professor Sune Visits the Other Bay

The island had a single low hill which rose smoothly from sea-level. The group skirted round it to reach a much larger bay on the far side. Here, a small city of tents was established in the dunes, with dozens of teams of anthropologists filming interviews with natives.

"In fact..." began the Knownot.

Again the Knave interrupted: "One can learn much from the Knownot's words. However, time does not press. I don't want to introduce you to the lead researcher. Professor Sune: this is Professor Moriarty. Professor Essor: this is Professor McGonagall."

Professor Essor, clearly well-versed in the niceties of Knavish introductions, replied smoothly: "Delighted to meet you, Professor Sune. Let me explain our work (Newheiser 2015). We are examining all possible collections of binary statements that Reception Committees may make. These are statements which can only be true or false, not indeterminate – equivalent to the concept of binary questions.

"Now there are only six possible assignments of the three tribes to the three Reception Committee members. Any statement by an islander can be classified according to which assignments it eliminates. For example, if a committee member says '*That* islander is a Knave!' pointing to one of the other committee members, then we can eliminate the possibility that a Knight is speaking about a Knownot. So in this particular case there are five possibilities which remain, out of the original six.

"A speaker can say nothing to prove he is not a Knownot. So two of the six assignments, those where the speaker is a Knownot, cannot be eliminated. But we may be able to exclude some of the other four assignments. Therefore there are just $2^4 = 16$ unique statement forms possible, based on which of the four eliminatable assignments are actually eliminated. Some statements may be made in ways which are verbally different but logically equivalent. For example: 'I'm no Knight!' and 'I'm a Knave!' appear to say different things but both only serve to prove that the speaker is the Knownot. So effectively, all binary statements fall into just these 16 forms. In a way it's curious that every possible binary statement, from 'It's sunny tonight!' to 'I love you more than apples!' collapses into one of these 16 forms. But I digress."

Sune nodded: "Often, statements depend on prior statements. A committee member may make a remark corresponding to one of your forms, and then another, fuming with indignation (feigned or real), may say 'That's not true!' One might think that's a new form, but really it isn't. Any new binary statement, in the context of statements previously made, can just serve to eliminate some of the remaining possible assignments, that's all. So it just behaves like one of your 16 existing forms."

Essor continued: "Indeed. Here, Professor Sune, have a look at this Truth Table. (Although why it's called a Truth Table escapes me. It shows both truths and untruths, so why does the name focus on half of the content? It's like calling a black-and-white photo a black photo. Or maybe it's meant to imply that the *content* of the table is accurate. But it's no more truthful than say a multiplication table. *Any* tableau of correct data should be true in that sense.)

"Anyway, again I digress. Suppose our Committee members are named A, B and C, and A makes a statement. The table shows examples of all the 16 possible forms and which of the 6 assignments they support. The column labelled TFR for example covers the case where A is a Knight, B is a Knave, and C is a Knownot. The first column shows how adjacent pairs of statements can be tied together as the alternative answers to binary questions directed to A. S01 and S02 for example are the two responses to question Q01, which might take the form: 'Are you a Knight?' The final column shows the total number of assignments which remain possible after the statement. Does that make sense, Professor Sune?"

"Yes," said Sune. "In fact, your first column lets me translate this into my world of binary questions. Suppose I ask Q01. If the response to Q01 is 'No!' (S02) then I am nearly home with a question to spare, because I have identified the Knownot already. Hurray! But if the response is 'Yes!' (S01) then I feel foolish because I have learned nothing from my question.

"In order to guarantee that I will finish within three questions, the first question I ask must be one which splits the possibilities evenly: four vs four. If not, there is a chance that we will still have more than four possibilities remaining with only two more questions to ask. So my first question must take the form of Q02, Q07 or Q08, which I see you have already marked in bold in your table. These correspond to asking A: 'Are you a Knownot?' 'Is B a Knownot?' or 'Does B lie more than C?'

"But, in fact there's an additional requirement. The first question *must* exclude one person from being the Knownot, or otherwise we cannot be sure that the second question will get us down to fewer than 3 possibilities. Only Q08, 'Does B lie more than C?', achieves this. So our first question *must* be of this general form (Tables 10.1 and 10.2)."

Table 10.1 Essor's truth table of possible statements by A

Q Id.	S Id.	Example	Comments	RTF	RFT	TRF	FRT	TFR	FTR	Total
Q01	S01	"I'm a Knight."	= "I'm no Knave" No information	Y	Y	Y	Y	Y	Y	6
	S02	"I'm no Knight."	= "I'm a Knave." Only Knownots can say it	Y	Y	N	N	N	N	2
Q02	**S03**	**"I'm no Knownot."**	**Knaves can't say it**	**Y**	**Y**	**Y**	**N**	**Y**	**N**	**4**
	S04	**"I'm a Knownot"**	**Knights can't say it**	**Y**	**Y**	**N**	**Y**	**N**	**Y**	**4**
Q03	S05	"B's a Knave"		Y	Y	N	Y	Y	Y	5
	S06	"B's no Knave"		Y	Y	Y	N	N	N	3
Q04	S07	"C's a Knave"		Y	Y	Y	Y	N	Y	5
	S08	"C's no Knave"		Y	Y	N	N	Y	N	3
Q05	S09	"B's a Knight"		Y	Y	N	Y	N	N	3
	S10	"B's no Knight"		Y	Y	Y	N	Y	Y	5
Q06	S11	"C's a Knight"		Y	Y	N	N	N	Y	3
	S12	"C's no Knight"		Y	Y	Y	Y	Y	N	5
Q07	**S13**	**"B's a Knownot"**	**= "C's no Knownot"**	**Y**	**Y**	**Y**	**N**	**N**	**Y**	**4**
	S14	**"B's no Knownot"**	**= "C's a Knownot"**	**Y**	**Y**	**N**	**Y**	**Y**	**N**	**4**
Q08	**S15**	**"B lies more than C"**	**= "If you asked me, I'd say B's a Knewnot"** [a]	**Y**	**Y**	**N**	**N**	**Y**	**Y**	**4**
	S16	**"C lies more than B"**	**= "If you asked me, I'd say B's a Knewnot"**	**Y**	**Y**	**Y**	**Y**	**N**	**N**	**4**

[a] We have to be careful about how a Knownot would interpret Essor's alternative phrasings for S15 and S16: "If you asked me, etc." Our Knownot considers each implied utterance within their remark and determines independently whether to lie or tell the truth about it. So to answer "If I asked you whether you like strawberries, would you say 'yes'?" would require two flips of an imaginary coin. For now we can use the simpler "B lies more than C" pattern.

"I agree," said Essor. "Now, to avoid drafting an entirely new table as we proceed, let's imagine that our first question had in fact been to B or C, and had eliminated the possibility that A is the Knownot. For the second question, we can then use the table that we had before, but eliminate the columns RFT and RTF. The sums are reduced too.

"We must pick a question that splits the remaining options two and two. The same three useful question forms marked in bold: Q02, Q07, Q08. After this question, there are two options to be distinguished by the final question that can be addressed to any

Table 10.2 Essor's truth table of possible 2nd statements by A, after 1st statement (by B or C) proves that A is not a Knownot

Q Id	S Id.	Example	TRF	FRT	TFR	FTR	Total
Q01	S01	"I'm a Knight."	Y	Y	Y	Y	4
	S02	"I'm no Knight."	N	N	N	N	0
Q02	**S03**	**"I'm no Knownot"**	**Y**	**N**	**Y**	**N**	**2**
	S04	**"I'm a Knownot"**	**N**	**Y**	**N**	**Y**	**2**
Q03	S05	"B's a Knave"	N	Y	Y	Y	3
	S06	"B's no Knave"	Y	N	N	N	1
Q04	S07	"C's a Knave"	Y	Y	N	Y	3
	S08	"C's no Knave"	N	N	Y	N	1
Q05	S09	"B's a Knight"	N	Y	N	N	1
	S10	"B's no Knight"	Y	N	Y	Y	3
Q06	S11	"C's a Knight"	N	N	N	Y	1
	S12	"C's no Knight"	Y	Y	Y	N	3
Q07	**S13**	**"B's a Knownot"**	**Y**	**N**	**N**	**Y**	**2**
	S14	**"B's no Knownot"**	**N**	**Y**	**Y**	**N**	**2**
Q08	**S15**	**"B lies more than C"**	**N**	**N**	**Y**	**Y**	**2**
	S16	**"C lies more than B"**	**Y**	**Y**	**N**	**N**	**2**

known un-Knownot (e.g. the second questionee). No need for a table to show that third question.

"However in my own work, I do not question islanders: instead I indicate that they are each free to make any one statement of the sixteen different forms. So there are basically $16^3 = 4096$ possible situations. Allowing for symmetries, this reduces to 816 different possible situations. Fortunately, my University has good funding for this kind of work from a generous benefactor, and we have been able to conduct detailed experiments here on the island, combined with rigorous computer analysis to conclude that exactly 161 of these 816 situations are contradictory and cannot occur. Of the remaining 635 viable situations, exactly 274 have a unique solution. If I was minded to mine this valuable anthropological research to make logical puzzles, as *some* have done, I could hence publish exactly 274 different puzzles."

Sune responded: "I've seen websites (Ernst et al. 2015; Hafner 2013a, b, c) demonstrating Knights and Knaves puzzles generated and solved by computer. One at Hong Kong University shows hundreds of progressively harder puzzles involving up to nine Knights and Knaves (no Knownots), and involve more general question phrasings involving up to three islanders at a time. For that, there would be up to $2^{2^3} = 256$ different statement forms, and I don't know how many different puzzles might be made from that."

"Yes," said Essor. "I've also begun to dabble in looking at larger crowds of up to a hundred islanders. (Since there is no longer exactly one member per tribe, I don't use the term 'committee'.) What kind of strategies can determine all the Knownots in a crowd? An interesting point is that we must assume that the Knownots are in the minority. If the Knownots ever equal or exceed the number of Knights plus Knaves, then one could never determine uniquely who is from which tribe. Rather a frightening thought!"

Sune nodded: "Yes, I've read about that idea before (Buchanan and Conway 2015). But it's eclipsed by other work. The basic idea is that a conspiracy of Knownots within the crowd might agree that each should behave as a Knight or a Knave. And if these conspirators stick consistently to their role-play, and if there is enough of them, then there is no way you could disprove the hypothesis that everyone *outside* the conspiracy is a Knownot instead. So we must specify that the Knownots be a strict minority in any crowd."

Essor smiled. "Ah! But how can you could ensure this? How do you conduct interviews for crowd membership? The broader population from which the crowd is drawn might itself have a majority of Knownots. I hear that the leading crowd researcher, Professor Usely, is in fact sailing to this island now. So maybe he will explain his 'crowd-sourcing' approach."

10.5 In Which Sune Talks About Exploding God-Heads

"I look forward to his arrival," said Sune. "In the meantime, the question of one statement referring to the truth of another brings up another broader point. We can imagine a pair of statements:

A: B's next statement will be true.

B: A's last statement was true.

or indeed a single statement:

A: This statement is false.

"What is one to conclude from these? There is clearly some Funny Business afoot. Smullyan is so articulate in *What is the Name of This Book?* about this confusing area that I want to quote him at length. Let me see: p. 216."

She fished out a dog-eared copy from her pack, and read:

> Now, what is wrong with the reasoning in these paradoxes? Well, the matter is subtle and somewhat controversial. There are those (philosophers, interestingly enough, rather than mathematicians) who rule out as legitimate any sentence which refers to itself. Frankly, I see this point of view as utter nonsense! In a self-referential sentence such as, "This sentence has five words," the meaning seems as clear and unequivocal as can be; just count the words and you will see the sentence must be true. Also, the sentence, "This sentence has six words," though false, is perfectly clear as to its meaning - it states that it has six words, which as a matter of fact it does not have. But there is no doubt about what the sentence says.

> On the other hand, consider the following sentence:
>
> THIS SENTENCE IS TRUE
>
> Now the above sentence does not give rise to any paradox; no logical contradiction results from assuming the sentence to be true or from assuming the sentence to be false. Nevertheless, the sentence has no meaning whatsoever for the following reasons:
>
> Our guiding principle is that to understand what it means for a sentence to be true, we must first understand the meaning of the sentence itself. For example, let X be the sentence: Two plus two equals four. [...] This illustrates what I mean when I say that the meaning of a sentence X being true is ***dependent*** on the meaning of X itself. If X should be of such a peculiar character that the very meaning of X ***depends*** *on the meaning of X being true, then we have a genuinely circular deadlock. [...] Sentences having this feature are technically known as sentences which are* **not well-grounded**.

"Similarly, the pair of statements by A and B above, considered as a system, are not well-grounded. It's not the predictive nature of A's remark which gives us the logical trouble. A might say this in a system which is perfectly well-grounded, e.g.:

> A: B's next statement will be true.
>
> B: I like strawberries.

"In the world of binary questions, we are only allowed to ask questions that have two possible answers, for any state of the world. If there was a state of the world where the respondent might get into a twist, unable to answer 'yes', and unable to answer 'no', then that uncertainty would constitute a third answer, breaking the rule. Brian Rabern and Landon Rabern (in their work on the so-called 'Hardest Logic Puzzle Ever') (Rabern and Rabern 2008; Bellos's 2016), **do** merrily violate that rule, while ensuring that questions are still well-grounded.

"They imagine three gods (True, False and Random) whose heads explode if they can't respond verbally to a question. It's unclear whether the gods recover to allow subsequent questioning, and if they do, what memories they have of their prior ordeal. They cannot recall all, or otherwise whenever a logician approached, they would run a mile, holding their fragile crania and shouting: 'Argh! I've met you lot before! Leave me alone!'.

"The Raberns' basic scenario is based on one from Smullyan via Boolos (1996). It is similar to our Knight, Knave and Knownot world, but there are differences. One additional wrinkle in Boolos and Rabern is that the local words for 'yes' and 'no' are 'da' and 'ja', and apparently the anthropologists don't know which is which.

"However a more significant novelty is Boolos' Random god. Basically, Random flips an imaginary coin once per utterance to decide whether to be a Knight or a Knave *for the entire duration of that statement*. Further, Random has self-knowledge about this internal state. So he can reply to embedded questions like 'If I asked you in your current state whether you like strawberries would you say "ja"?'

"The Raberns lobotomize Random to become just an imaginary coin-flipper and utterer of 'ja' and 'da', without any insight as to the truth of what it says. This may seem a sad state, but it least it renders him immune to Exploding Head Syndrome. So there is that consolation.

"The Raberns' basic solution involves the concept of embedded questions. This is a powerful idea introduced by Smullyan, and used widely in his books, to cut through separable layers of Knight and Knaves, 'um's and 'er's, sanity and insanity, etc. Let E be the function which takes a question, q, to the question 'If I asked you "q", would you say "ja"?', When any non-Random god is asked E(q), a response of 'ja'/'da' indicates that the correct answer to q is 'yes'/'no' respectively. Using this we can ask god A E('Is B Random?') which is just a camouflaged version of the essential "Which of the other two lies more?" question form.

"However, the Raberns also noted that there were no restrictions placed by Boolos that the questions be binary, I am afraid they could not resist the temptation to engage in some funny business." She shook her head in regret. "They asked the gods "Are you going to reply 'ja' to this question?" As they recorded gleefully in their expedition journal:

> If 'ja' means 'no' then True will be unable to respond with the truth. If 'ja means yes, then False will be unable to respond with a lie. But they are infallible gods! They have but one recourse – their heads explode

"As they wiped the ichor from their faces, I've no doubt Rabern, and the equally merry Rabern, reflected heartlessly that with this third outcome to one question, it may be possible to identify the three gods with just two questions in Boolos' original world, since $3 \times 2 = 6$. The key is a recursive first question. They call it Query-1, and it goes: E('Is it the case that: [(in your current mental state you would always answer "da" to Query-1) AND (B is True)] or (B is false)?')."

10.6 In Which Professor Sune Recounts Another Adventure

Essor invited Sune to dine with his team that night. Over dinner, she said: "Here is a simple problem I encountered a few months ago, which you might find interesting.[6] I was visiting a distant island inhabited by a single tribe. I understood nothing of the local dialect, although all the islanders can speak English. I had heard that the inhabitants will always give a truthful answer if they possibly can, but will answer randomly if they cannot give a truthful answer.[7] I knew that their words for 'yes' and 'no' were 'er' and 'um', but not which was which. I could ask one yes/no question which they might choose to answer either in English or in their dialect. Their answer should tell me the meanings of 'um' and 'er'. What could my question be?"

You might like to solve this challenge before reading on.

---

[6]Questions posed by Simon Norton in an email 2005 (Norton 2005). He said that he had presented the problem to Smullyan at one point, but we are not sure if it was circulated further.

[7]Being random if one cannot be truthful, is an interesting principle, but not actually used in the solution. If Knights, Knaves and Knownots were to adopt this principle, a wider range of questions than before could be regarded as binary. For example, we could ask "Are you uttering 'no'?" We wouldn't get any sense from anyone, but at least we would get a binary response.

"Suppose I ask that old standby: 'Does $1 + 1 = 2$?' Well, I might get a useful answer in the local language, but there is the possibility that I would get the answer 'Yes!' In which case, I have learned nothing.

"Or maybe I blithely ask: 'Does "um" mean "no"?' If someone responded in English, then I would have all the information I need. It was an island of truth-tellers, after all. But if the response was 'er', I would be no further forward.

"Some kind of trickery is inescapable. The question has to make some reference to the content of the answer. Otherwise, there are basically two situations we may face: either (1) it's the kind of question where the correct answer doesn't vary with the assignation of 'um'/'er' to 'yes'/'no', or (2) it does vary. There's no mystery about this: just look at the question and see which category it falls into.

"If the correct answer doesn't vary with the assignation (such as 'Does $1 + 1 = 2$?'), then if I get that answer in English I know nothing, as the answer is consistent with either assignment. If the correct answer *does* vary with the assignation (such as 'Does "um" mean no?'), then I may get the answer in the local dialect, in which case again I have nothing (Table 10.3).

"But can our funny business be well-grounded in Smullyan's sense? I wondered about asking them a circular question like: 'Are you uttering "um"?' It's contrived, but no odder than a statement like "This statement has five words." which we discussed earlier and deemed acceptable. However the answer 'no' is always valid, and doesn't help us to understand the meaning of 'um'.

"The next suggestion was: 'Are you uttering "no"?' Eureka! Respondents could not reply truthfully in English. So they must reply in their own language, and whichever word they used means 'no'.[8]

"A more sophisticated challenge is to pose a question to learn the answer to a particular random yes/no question (RQ)."

Again, you might like to solve this challenge before reading on.

"Note that RQ might be "Does 'um' mean 'no'?" so this challenge is a generalization of the previous one. It follows that trickery still reigns. The obvious approach to try would be an embedded question. The form could be: "(If you are answering in island dialect, then does 'um' mean 'yes') if and only if RQ? Let's see how this works out with a Truth Table (Table 10.4)."

Sune continued: "Let's go through the columns in turn. The first three show the input variables. What is the answer to RQ? What's the meaning of "um"? And how does the islander answer? There are $2 \times 2 \times 4 = 16$ possible cases. In column 4, we evaluate what is the correct answer to the If Question (IQ): 'If you are answering in island dialect, then does "um" mean "yes"?' This is based on columns 2 and 3. Remember that an if statement evaluates to true in 75% of cases. In column 5, we compare the correct answers to RQ and IQ and determine what the Islander can legitimately say (bearing in mind the meaning of 'um'). So for example in the first row, the answers to RQ and IQ are both the same ('yes'), and so the Islander can

[8]We think the basic puzzle can also work well as a dinner party game. Those who still feel psychologically scarred from the Reception Committee problem at the previous dinner party can be reassured that with *this* puzzle there are only truth-tellers. So it must be simpler, surely?

Table 10.3 (If you are answering in island dialect, then does "um" mean "yes") IFF RQ

RQ correct answer	Does "um" mean "yes"?	Islander's answer?	Correct answer to IQ	Possible consistent answers	Was real answer consistent?
Yes	Yes	Yes	Yes	Yes, um	Y
Yes	Yes	No	Yes	Yes, um	N
Yes	Yes	um	Yes	Yes, um	Y
Yes	Yes	er	Yes	Yes, um	N
Yes	No	Yes	Yes	Yes, er	Y
Yes	No	No	Yes	Yes, er	N
Yes	No	Um	No	No, um	Y
Yes	No	Er	No	No, um	N
No	Yes	Yes	Yes	No, er	N
No	Yes	No	Yes	No, er	Y
No	Yes	Um	Yes	No, er	N
No	Yes	Er	Yes	No, er	Y
No	No	Yes	Yes	No, um	N
No	No	No	Yes	No, um	Y
No	No	Um	No	Yes, er	N
No	No	Er	No	Yes, er	Y

Table 10.4 Consistent rows from Table 10.3

RQ correct answer	Does "um" mean "yes"?	Islander's answer?	Correct answer to IQ	Possible consistent answers	Was real answer consistent?
Yes	Yes	Yes	Yes	Yes, um	Y
Yes	Yes	Um	Yes	Yes, um	Y
Yes	No	Yes	Yes	Yes, er	Y
Yes	No	Um	No	No, um	Y
No	Yes	No	Yes	No, er	Y
No	Yes	Er	Yes	No, er	Y
No	No	No	Yes	No, um	Y
No	No	Er	No	Yes, er	Y

say something that means 'yes'. In this case 'yes' or 'um' are acceptable. Finally in column 6, we see whether what the Islander actually said in column 3 appears in the list of legitimate responses in column 5. If so/not, it is consistent/inconsistent, and is marked Y/N. Exactly half of the rows are inconsistent, so can't happen. Let's draw a new table with those filtered out.

Sune wrapped up: "Now it's easy to see that if the correct answer to RQ is 'yes' (column 1), the islander will say 'yes' or 'um' (column 3). If the correct answer is 'no', the islander will say 'no' or 'er'. We don't get any insight from this process as to whether 'um' means 'yes' - unless RQ probes this very point!"

10.7 Professor Usely Arrives

In the morning, Sune awoke. All was quiet. When she went out into the street, she found it was nearly deserted.

"Where's everyone gone?" she asked an old man.

"They've all gone to the moon!"

She tried someone else, who said: "There's a new researcher on the island. He needs just about everybody!"

Sune went back to the big bay. Beyond Essor's camp, a much more massive camp had appeared overnight, perhaps associated with the liner now anchored in the sea beyond. Hundreds of islanders were all gathered in a group. She found Essor.

"What's going on here?" she asked.

"Usely arrived late last night. He's trying to determine the number of questions necessary when there are more than 3 members in the group."

"How many members?"

"Hundreds! He's been conducting research on two other islands. On one of them, there were just Knights and Knaves. On the other, there were just Knights and Knownots. Curiously though, he lumps Knaves and Knownots together with the same terminology, calling them 'Spies', but in a different sense from Smullyan's. He's just interested in determining whether people are Knights or not. He only takes statements of the form 'Is so-and-so a Knight?' So now he's here to extend his research by bringing all three tribes together (Wildon 2015). Apart from the great anthropological interest in such work, there may also be implications for fault diagnosis in multi-processor systems. This work is basically combinatorial in focus, trying to see how logical reasoning can scale."

They walked across the beach to where a large man with a panama hat was engaged in directing the labours. Introductions were made.

"Thanks for the welcome," said Usely, "yes I've been looking at Knights and Knaves, and Knights and Knownots for many years now. I have determined accurate formulae for the number of questions of the form 'Person X: is person Y a Knight?' required to achieve the following tasks, where there are n people in a room, at least k of whom are knights, and $k > n/2$:

- Identify 1 Spy, or confirm no-one is.
- Identify 1 Spy, given that a spy is present.
- Find a Knight.
- Find the identity of at least one person.
- Find the identity of a particular person.
- Find the identity of a particular person, given that a Spy is present.

"Where the Spies are Knaves, it's generally possible to pursue these six tasks simultaneously. However where the Spies are Knownots, this is not possible – there are different optimal strategies. Quite interesting. I'm extending prior work by Blecher (1983) with Knights and Knownots. For a long time, $2n - k - 1$ questions were known to be necessary and sufficient. E.g. if $n = 7$ and $k = 4$, then 10 questions are necessary."

"Impressive," said Essor. "I had one question. How do you conduct interviews to ensure that there are not too many Spies in your groups?"

"For Knights and Knaves, it's not a problem. At the interviews we can just ask each of them if $1 + 1 = 2$. But for Knownots, there's no solution. It can only be an assumption that there are less than half Knownots. If more than half the world were Knownots, we could never prove it. This 'Knownot Apocalypse' is a scary thought!"

10.8 Wrap-Up, Open Questions

This article has been a sampler of the burgeoning field of "Knight-Knave studies", which runs the gamut from trivial fun to serious mathematics. To our minds, there are many directions still to be explored. Here are just a few:

(1) Wildon lists several open questions: one of the most interesting is to get a precise expression for the number of questions necessary to identify all identities in Knights and Knaves groups, with n individuals of which at least k are knights, $k > n/2$. This is not yet fully known.
(2) What is the best question strategy if we are not restricted to Usely's single question form?
(3) If we are only allowed one kind of question form (like in Sect. 10.1) what is the best form?
(4) Another way to extend Wildon's work is to do the study that "Usely" came to the island to perform. How do the search strategies scale when Knaves and Knownots are combined together with Knights?
(5) What other question forms are there involving a maximum of 4 individuals including the speaker? How does this impact the number of questions required?
(6) It's assumed that all the islanders know the tribe of all the other islanders. Suppose we limit the information known by islanders, so each is only aware who belongs to his own tribe. How does that affect the identification process?
(7) How about situations where the number of Knownots exceeds the number of Knights and Knaves? Clearly, it will be impossible to determine a unique answer, but what can one conclude?
(8) How about extending the use of embedded non-binary yes/no questions (and "ah!" rather than head-exploding) to cases with larger numbers of islanders including Knownots?
(9) How many different puzzles can exist with up to n Knights/Knaves [/Knownots], using the minimum number of statements that will serve to identify them

uniquely, where statements are permitted to mention up to m individuals apart from the speaker?

(10) The basic assumption made in these problems is a conservative one: we want to make sure that our questioning leads to a solution in all circumstances. Suppose we take a probabilistic approach, just looking for the best expectation of success, what then?

(11) A more systematic exploration of the "middle tribe" could be fruitful. Three behaviours were identified in the tale: Knownots, Boolos' Random and Raberns' Random. Can you distinguish them on an island with no Knights or Knaves?

(12) What kind of information can be gleaned from questions: "Jimmy, did you eat the cake?" or imperative statements: "Go to your room, Jimmy!" What implications are there about the existence of Jimmy, cake and room, if a Knave makes these utterances? What scope for deceit does a Knave or a Knownot have?

(13) Suppose Professor Sune bravely sails her little schooner next to a realm where classical logic no longer applies. Can we explore Knights and Knaves with non-classical logics (Rosenhouse 2014)?

(14) A final challenge for eager young anthropologists: "At its core, quantum mechanics can be regarded as a non-classical probability calculus resting upon a non-classical propositional logic (Wilce 2002)." Can some trickster compose a quantum mechanical Knights and Knaves puzzle?

So thanks to Raymond Smullyan, for endless hours of logical amusement, still unending.

References

Raymond M. Smullyan (1978) *What Is the Name of This Book? The Riddle of Dracula and Other Logical Puzzles* Prentice-Hall

Raymond M. Smullyan (1982a) *The Lady or the Tiger?* Knopf

Raymond M. Smullyan (1982b) *Alice in Puzzle-Land* Morrow

Raymond M. Smullyan (1985) *To Mock a Mockingbird* Knopf

Raymond M. Smullyan (1987) *Forever Undecided* Knopf

Raymond M. Smullyan (1992) *Satan, Cantor and Infinity* OUP

Raymond M. Smullyan (1997) *The Riddle of Scheherazade* Knopf

Raymond M. Smullyan (2007) *The Magic Garden of George B. And Other Logic Puzzles* Polimetrica

Raymond M. Smullyan (2008) *Logical Labyrinths* CRC Press

Raymond M. Smullyan (2013) *The Godelian Puzzle Book: Puzzles, Paradoxes and Proofs* Dover

Martin Gardner (2013) *My Best Mathematical Games and Puzzles* pp 2,40 Courier Corporation

Werner Herzog (1974) *The Enigma of Kaspar Hauser*, movie

Simon Norton (2015) personal communication

Mark Newheiser (2009-2015) *Ordinary Logic Taken to Illogical Extremes*, sites.google.com/site/newheiser/knightsandknaves

Joe Lau, Jonathan Chan, Zachary Ernst (2004-2015) *Knights and Knaves Puzzles* http://philosophy.hku.hk/think/logic/knights.php

Isidor Hafner (2013a) *Knights and Knaves Puzzle Generator – Wolfram Demonstrations Project* http://demonstrations.wolfram.com/KnightsAndKnavesPuzzleGenerator/

Isidor Hafner (2013b) *Another Knights and Knaves Puzzle Generator – Wolfram Demonstrations Project* http://demonstrations.wolfram.com/AnotherKnightsAndKnavesPuzzleGenerator/
Isidor Hafner (2013c) *Knights, Knaves, and Normals Puzzle Generator – Wolfram Demonstrations Project* http://demonstrations.wolfram.com/KnightsKnavesAndNormalsPuzzleGenerator/
Buchanan, Andrew G., & Conway, John H. (2015). Anthropologic. *The Mathematical Gazette*, *99*(546), 537–538.
Rabern, B., & Rabern, L. (2008). A simple solution to the hardest logic puzzle ever. *Analysis*, *68*(2), 105–112.
Alex Bellos's Monday puzzle (2016, 29 Feb) The Guardian https://www.theguardian.com/science/2016/feb/29/can-you-solve-it-the-puzzee-rascal-problem-that-will-drive-you-bonkers
Boolos G. (1996) The hardest logic puzzle ever. *The Harvard Review of Philosophy 6*, 62–65. Reprinted in his *Logic, Logic, and Logic*, 406–410. 1998. Cambridge, Mass.: Harvard University Press
Simon Norton (2005) personal communication
Mark Wildon (2015, Apr) Searching for knights and spies: a majority/minority game, http://www.ma.rhul.ac.uk/~uvah099/Maths/MinorityGame21.pdf
Blecher, Pavel M. (1983). On a logical problem. *Discrete Math.*, *43*, 107–110.
Rosenhouse, Jason. (2014). Knights. *Knaves, Normals, and Neutrals, The College Mathematics Journal*, *45*(04), 297–306.
Alexander Wilce (Feb 2002) *Quantum Logic and Probability Theory* Stanford Encyclopedia of Philosophy http://plato.stanford.edu/entries/qt-quantlog/

Chapter 11
Making The 'Hardest Logic Puzzle Ever' a Bit Harder

Walter Carnielli

Abstract This paper intends to propose new forms of logic puzzles by adopting a pluralist perspective. Not only can this expanded view lead to more challenging puzzles, but it also helps the understanding of novel forms of reasoning. In 1996, George Boolos published a famous puzzle, known as the 'hardest logic puzzle ever'. This puzzle has been modified several times, and is known not to be 'the most difficult of all logical puzzles'. I argue that modified versions of this famous puzzle can be made even harder by using non-standard logics. As a study case, I introduce a version of the puzzle based on the three-valued paraconsistent logic **LFI1** and show how it can be solved in three questions, leaving the conjecture that this three-valued puzzle cannot be solved in fewer than three questions.

11.1 Harder and Harder Puzzles

Almost a quarter of a century ago, George Boolos published a short paper (Boolos 1996) that he qualified as 'the most difficult of all logical puzzles', attributed to the logico-mathemagician Raymond Smullyan and modified by the computer scientist John McCarthy, although related puzzles can be found throughout Smullyan's books since the seventies.

More than just another logical puzzle involving gods, genies, and strange tribesmen who always tell the truth or lie, Boolos' puzzle possesses interesting peculiarities that make it highly non-trivial (though not to the maximum degree, as we shall see) and philosophically relevant. Boolos' original puzzle is as follows: Three gods A, B, and C are called, in some order, Verus, Falsus and Aleatorius. Their English nicknames are True, False and Random. Verus always tells the truth, Falsus always lies, and Aleatorius always speaks truthfully or lies in a completely random way. Your task is to determine, in the original version of the puzzle, the identities of A, B, and C with three questions whose answers are 'yes or no'; each question should be posed

W. Carnielli (✉)
Department of Philosophy and Centre for Logic Epistemology and the History of Science, University of Campinas – UNICAMP, Campinas, SP, Brazil
e-mail: walter.carnielli@cle.unicamp.br

M. Fitting and B. Rayman (eds.), *Raymond Smullyan on Self Reference*, Outstanding Contributions to Logic 14, https://doi.org/10.1007/978-3-319-68732-2_11

to one of the gods. It may be that a god is asked more than one question or that a god is not questioned at all.

The gods understand your language,[1] but respond to all questions in their native language in which words for 'yes' and 'no' are *da* and *ja*, in some order. You do not know which of the words means 'yes' or 'no'. Boolos takes all necessary care to make the puzzle solvable and hard, but unfortunately not sufficient care – it has been shown in (Rabern and Rabern 2008) that the puzzle becomes almost trivial: indeed, Boolos on page 96 of Boolos (1996) says:

> Whether Random speaks truly or not should be thought of as depending on the flip of a coin hidden in his brain: if the coin comes down heads, he speaks truly; if tails, falsely.

However, this little *faux pas* makes Aleatorius be, after each question, randomly transformed into Verus or Falsus, who are semantically committed (it is not easy to be a constant truth-teller or a perpetual liar), and in this way Aleatorius is not a true random replier. The puzzle has been conveniently amended in Rabern and Rabern (2008) by changing Boolos' original condition to:

> Whether Random answers *ja* or *da* should be thought of as depending on the flip of a coin hidden in his brain: if the coin comes down heads, he answers *ja*; if tails, he answers *da*.

This amended version makes Aleatorius truly random by avoiding any reference to semantics and thus avoiding trivialization, and is shown to be solvable in three questions. A solution in two questions to the amended puzzle is shown in Uzquiano (2010), and a further modification of the original puzzle, eliminating part of the available information, is solved in Novozhilov (2012). More on the unfoldings around this puzzle follows.

I will discuss some ways to make the 'hardest puzzle ever' even harder, or at least more involved, by taking seriously the role of the logic behind the gods.

11.2 Gods Reasoning with Enhanced Logic

Gods are, by definition, omniscient. They know what we humans know, and what we do not know or have doubts about. Verus and Falsus must be careful when answering questions like "Are humans causing global warming?", or even "Is the Goldbach conjecture true?", or (given any question q), "Is Aleatorius going to reply with no to q?".

Gods, being omniscient, know that certain facts that we believe to be absolutely correct may become dubious. The earth was once believed to be flat; in the nineteenth century Vulcan was believed to be a hypothetical planet orbiting between Mercury and the Sun, and so on. So what you *believe* should not be taken as absolutely certain, although what you *know* (either by proof, or as accepted scientific fact) can be taken as answerable in a definite yes-no way.

---

[1] Whatever it is, not only English: gods are not monolingual.

The first question above might be dubious for human knowledge if there is contradictory evidence about global temperature at the present stage, and the second is simply unknown for mankind despite much positive evidence. The third question, however, is inherently doubtful, even for our gods – in our logic below, questions of this sort are *inconsistent*, or *non-consistent*, in contrast to *consistent* questions such as "Does $0 = 1$?".

Besides that, omniscient gods are aware (as are some logicians and philosophers) that human belief is often challenged by contradictory evidence, that it is possible to build logics where one can reason about contradictoriness, and that nobody's head needs to explode in the presence of a contradiction if one embraces a careful way of reasoning such as a paraconsistent logic. Dealing with non-explosive contradictions may be something new for some logicians, but certainly not to gods or theologians[2] – solutions blaming the "incapacity" of logic only show a reluctance to take logic or divinity seriously.

So the gods, being aware of the problem of contradictory evidence and information for mankind, and that paraconsistent logics are known to people and that are even applicable to database models, will use one such logic to reply. Let us clarify some guidelines before proceeding:

G1 The gods know precisely what you know, and what you do not know or are doubtful about. If a question involves an absolutely undoubtful topic, for instance, "Is $0 \neq 1$ true?", Verus and Falsus will reply "affirmative" or "negative", but if it involves a doubtful topic that may admit contradictory answers, or has no answer at all, they will say "indeterminate". The gods understand any language, but will answer all questions in their own language, in which the words for "affirmative", "negative" and "indeterminate" are *da*, *ta*, and *ja*, in some order, but you don't know which word means which.[3]

G2 In this way, Verus and Falsus will, in this 3-valued enhanced puzzle, respond with three truth values, namely "affirmative", "negative" and "indeterminate", reserving "affirmative" and "negative" for definite (i.e., consistent) questions in specific domains, as for instance about Arithmetic,[4] abbreviated as 1, 0 and 1/2 , and Aleatorius will continue to reply randomly, now rolling a die hidden in his brain: if the die rolls 1 or 6, he says *da*; if 2 or 5 he says *ja*, and if 3 or 4 he says *ta*.

G3 Responding with three truth-values requires, of course, a modification of basic logic; the gods have agreed to use a simple expansion of propositional logic that does not betray any classical principle, but simultaneously permits one to extract information from contradictions and to reason with such contradictions – they are using the logic called **LFI1** described below (see Carnielli et al. 2000).

---

[2]This is seriously discussed e.g. in Rauser (2002).

[3]Just by coincidence, *da*, *ja* and *ta* correspond to *yes* in, respectively, Russian, German and colloquial Portuguese.

[4]It is debatable whether the a priori validity of Arithmetic and the impossibility of its empirical falsification can be demonstrated, but for the purpose of this puzzle let us honor this idea which Kant had already defended in the 18th century.

LFI1 represents a quite ubiquitious way to reason under contradictions without the risk of falling into trivialism. It coincides with da Costa and D'Ottaviano's logic **J3** of 1970, and with Batens and De Clercq's logic **CLuNs** of 2004. It also coincides with the logic Φ_v, a three-valued logic introduced in 1960 by K. Schütte in the context of Proof Theory.

Moreover, Lukasiewicz's 3-valued logic **L3** and **LFI1** are functionally equivalent, in the sense that all connectives of one can be defined in the other (but with truth values working differently: in **L3** only 1 is a distinguished truth value, while in **LFI1** both 1 and 1/2 are distinguished). In its turn, the tables of Lukasiewicz's **L3** for conjunction, disjunction, and negation coincide with Kleene's **K3** and Asenjo-Priest's logic of paradox **LP** of 1979.

So **LFI1** does not represent any frivolous invention: it is a strong fragment of classical logic, in the sense that any extension obtained by adding a previously rejected classical tautology to it will collapse into classical logic. **LFI1** is also a quite natural way of reasoning, as it has been rediscovered many times, and better exploits the distinction between contradictions and inconsistencies. It just seems natural that the gods choose **LFI1**, and they intend to reappear with other puzzles based on other non-standard logics.

The language of our logic is defined in the usual way, with the addition of a new unary connective $\bullet$ (read as "it is inconsistent"). So, all syntactical notions are familiar ones, with obvious modifications.

The deductive machinery of **LFI1** offers a formalization of what an inconsistency should be from the point of view of classical logic.[5] It is worthwhile to note that the concepts of contradiction and inconsistency are not necessarily to be identified (see e.g. Carnielli et al. 2007). But in the logic **LFI1** inconsistency and contradiction are coextensive, and it can be easily proved that inconsistent information (formalized as $\bullet A$) and contradictory information (in the form $A \wedge \neg A$) are identified through the equivalence $\bullet A \leftrightarrow A \wedge \neg A$.

The logic **LFI1** is defined by the following matrices, where 1 and 1/2 are distinguished truth-values:

$\vee$	1	$^1/_2$	0
1	1	1	1
$^1/_2$	1	$^1/_2$	$^1/_2$
0	1	$^1/_2$	0

	$\neg$	$\bullet$
1	0	0
$^1/_2$	$^1/_2$	1
0	1	0

Conjunction $A \wedge B$, implication $A \to B$, and consistency $\circ A$ are defined, respectively, as $\neg(A \vee \neg B)$, $\neg(A \vee \bullet A) \vee B$. and $\neg \bullet A$:

[5]Logics of this family are called Logics of Formal Inconsistency (LFIs) and are thoroughly discussed in Carnielli et al. (2007). A full account of such logics is found in Carnielli and Coniglio (2016).

$\wedge$	1	$1/2$	0
1	1	$1/2$	0
$1/2$	$1/2$	$1/2$	0
0	0	0	0

$\to$	1	$1/2$	0
1	1	$1/2$	0
$1/2$	1	$1/2$	0
0	1	1	1

	$\circ$
1	1
$1/2$	0
0	1

.

If A is any question about answers of questions posed to the gods, then $\circ A$ is true, because the answers are never themselves indeterminate (whether the content of the reply, as expressed in terms of da, ja and ta, is true, false, or contradictory is another story).

LFI1 is axiomatized and shown to be complete with respect to such three-valued semantics, or equivalently with respect to the following inductively defined binary interpretation:

1. $I \models A \wedge B \Leftrightarrow I \models A$ and $I \models B$
2. $I \models A \vee B \Leftrightarrow I \models A$ or $I \models B$
3. $I \models A \to B \Leftrightarrow I \not\models A$ or $I \models B$
4. $I \models \neg\neg A \Leftrightarrow I \models A$
5. $I \models \bullet A \Leftrightarrow I \models A \wedge \neg A$
6. $I \models \neg \bullet A \Leftrightarrow I \not\models \bullet A$
7. $I \not\models \neg A \Rightarrow I \models A$
8. $I \models \neg(A \wedge B) \Leftrightarrow I \models \neg A$ or $I \models \neg B$
9. $I \models \neg(A \vee B) \Leftrightarrow I \models \neg A$ and $I \models \neg B$
10. $I \models \neg(A \to B) \Leftrightarrow I \models A$ and $I \models \neg B$.

The binary semantics may seem more intuitive, but you lose truth-functionality (see clause (7) above). A solution using the three-valued characteristics of such a logic can be obtained in three questions.

Towards the solution, note first that Verus could reply 1/2 to question q: "Is there life elsewhere in the universe?", but if you asked, after this question, "Did you reply 1/2 to question q?" he would say 1. Falsus would also reply 1/2 to the same question, because "to lie" is to produce the negation of the correct truth value, and in **LFI1** negation of 1/2 is 1/2, but he would reply 0 to the next question, "Did you reply 1/2 to question q?". This makes clear that, even if a question q may be doubtful, Verus and Falsus reply definitely 0, 1/2 or 1 to it, and questions about their answer are responded with 0 or 1, as a consequence of the above guideline (G2). Aleatorius, of course, will keep replying randomly. In the language of **LFI1** a question A of the form "Did you reply 1/2 to question q?" is consistent, that is, $\circ A$ is true, even if A may be false.

Let E, as in Rabern and Rabern (2008), be the function that takes any question q to the counterfactual question $E(q)$: "If I asked you q would you say *ja*?". The idea, credited to R. Smullyan for solving his own puzzle 162 in Smullyan (1978), can be conveniently generalized and is the core of the "embedded question lemma" below.

The lemma takes advantage of an essential feature of **LFI1**, namely the distinction between consistent and inconsistent questions and its effect on the underlying three-valued logic.

Lemma 11.2.1 *(Embedded question lemma, three-valued form)*

(1) When either Verus or Falsus are asked $E(q)$, a response ja implies that the correct answer to q is not negative (i.e., either 1 *or* 1/2*).*
(2) When either Verus or Falsus are asked $E(q)$, a response of any other values (da or ta) to consistent questions implies that the correct answer to q is negative.
(3) The responses to $E(q)$ exclude one of the values among ja, da, or ta. The excluded value corresponds to indeterminate.

Proof (1) Suppose, by *reductio ad absurdum*, that q is any question with a (definite) negative answer. Then it is impossible that $E(q)$ can be answered by *ja*, whatever the meaning of *ja*.

(2) Suppose, again by *reductio ad absurdum*, that q is any consistent question with a (definite) positive answer. Then it is impossible that $E(q)$ can be answered by anything different from *ja*, whatever the meaning of *ja*.

It is perhaps more intuitive, however, to directly compute all possible values involved. Let V, F represent Verus and Falsus, q a question, and $E(q)$ be the embedded question: "If I asked you q would you say *ja*?" Let us compute all possible values of q and *ja* in order to determine which answers to $E(q)$ are *ja*.

	q=0	Answers to	
		q	E(q)
ja=0	V	0	1
	F	1	1
ja=1	V	0	0
	F	1	0

	q=1	Answers to	
		q	E(q)
ja=0	V	1	(0)
	F	0	(0)
ja=1	V	1	(1)
	F	0	(1)

	q=1/2	Answers to	
		q	E(q)
ja=0	V	1/2	(0)
	F	1/2	1
ja=1	V	1/2	0
	F	1/2	(1)
ja=1/2	V	1/2	1
	F	1/2	0

1. Suppose q is definitely false, such as, e.g., "Does $0 = 1$?". Then no answer *ja* to q will be 1/2, since q is consistent, and consequently *ja* is either 0 or 1. In this case, for each choice of *ja*, there is no answer to $E(q)$ as *ja* .
2. Now suppose q is definitely true, such as, e,g., "Does $0 = 0$?". Then, again, no answer *ja* to q will be 1/2, since q is consistent, and consequently *ja* is either 0 or 1. In this case, for each choice of *ja*, the answers to $E(q)$ as *ja* are circled:
3. In the last case, suppose q is indeterminate, as e.g. "Is Aleatorius going to reply 0 to the next question q I will ask him?". Note that, since the right answer to q is 1/2, F replies with the negation of 1/2, which is also 1/2. In this last case, for each choice of *ja*, the answers to $E(q)$ as *ja* are circled.

   For (1) and (2), it is obviously clear, by inspection, that when either True or False are asked $E(q)$, a response *ja* (whichever it is) implies that the correct answer to q is either 1 or 1/2. It is also clear, from the first and second tables, that when either Verus or Falsus are asked $E(q)$, a response distinct of *ja* to consistent questions implies that the correct answer to q is negative.

   (3) It follows from the fact that questions of type $E(q)$ can never be answered indeterminately, by guideline (G2) (more visually, the responses to $E(q)$ in the tables exclude one of the values among *ja*, *da* or *ta*, and the excluded value corresponds to the indeterminate truth value 1/2).

□

Note that here we cannot guarantee, as in other versions of the puzzle, that a response of *da* or *ta* indicates that the correct answer to q is negative for any question, just for the consistent ones.[6] It is also instructive to note that the metalogic (that is, the logical environment where proofs are carried out, discussions are maintained, etc.) is classical—not only in this puzzle, but in almost all literature in logic. Only the logic of the gods is non-classical.

What the previous lemma shows is that suitable counterfactual conditional questions may get a lot of information about gods and truth values. The remaining problem then is to separate Aleatorius from the others, a task solvable in two questions as in other versions of the puzzle. How can this be done? Again, with specific consistent counterfactual conditional questions of the type $E(q)$. For example, with no loss of generality, ask B, listing the gods in some linear order:

(a) If I asked you 'is A Aleatorius?' would you answer *ja*?

Case (1) If B is Aleatorius, then A and C are not Aleatorius.
Case (2) If B is either Verus or Falsus, then:

(2.1) If B answers *ja*, by Lemma 11.2.1(1), since the question 'is A Aleatorius?' is consistent (that is, it admits no answer 1/2), the correct answer is affirmative, and indeed A is Aleatorius. So C is not Aleatorius.

[6] Clearly, a function $E'(q)$: "If I asked you 'q would you say x?", where x is *ja*, *da* or *ta* produces a similar effect, in the sense that a response of x indicates that the correct answer to q is affirmative, but $E(q)$ suffices to the solution.

(2.2) If B replies with any other value, say (with no less of generality) *ta*, by Lemma 11.2.1(2) the correct answer in this case is negative, so *A* is not Aleatorius.

Hence, in all cases, after one question we know at least the identity of one god who is certainly not Aleatorius. Suppose (again with no loss of generality) this is *C*. Now ask *C* the following two questions:

(b) "If I asked you 'is *C* Verus?', would you answer *ja*?"
(c) "If I asked you 'is *B* Aleatorius?', would you answer *ja*?"

Again, by Lemma 11.2.1(1) and (2), as *C* is Verus or Falsus, question (b) determines his identity, and question (c) determines whether *B* is Aleatorius or not; hence the identity of *A* is known.

So this is a simple solution in three questions, which works only if we ask appropriate consistent questions and understand how the logic **LFI1** is built. I conjecture that this three-valued puzzle cannot be solved in fewer than three questions, but don't know how to prove it.

11.3 Closing

In Uzquiano (2010) a two-question solution is offered to the amended "classical" puzzle, but I do not know whether that solution can be adapted to this expanded puzzle. I conjecture that it is impossible to solve with fewer than three questions. A harder version is proposed in Uzquiano (2010), argued to be unsolvable in fewer than three questions by Wheeler and Barahona (2012) (by means of an information theoretical argument that considers a three-valued answering repertoire). However, Wintein (2012) is able to bypass this conclusion by employing a four-valued answering repertoire. Although this is an interesting use of expanded, non bi-valued, scenarios and advanced theories of truth as tools for solving puzzles (see also Wintein 2011), what I introduce here is the idea of proposing new puzzles based upon non-classical logics, by explicitly widening the logic behind the puzzle agents. However, the intuition of making puzzles harder by allowing the gods to answer through a many-valued repertoire is already present in Wintein (2012).

A formalism to implement automatic model-checking techniques based on public announcement logic is introduced in Liu and Wang (2012). Their setting is handy for proving impossibility results, as non-existence of solutions to certain harder puzzles, but only treats varieties of standard puzzles.

As argued in Rabern and Rabern (2008) a simpler solution can be given to their amended puzzle, taking advantage of the limits of logic: suppose we ask one of the gods this question:

Are you going to answer *ja* to this question?

If *ja* means no, then Verus will be unable to respond with the truth. If *ja* means yes, then Falsus will be unable to respond with a lie. By skillfully tailoring questions of this sort, two questions will solve the problem. But precisely because they are gods, who are aware of the freedom of logic, they may rely on another logic. The use of **LFI1** is just one possibility: in this case, in order to explode, the question would have to refer to its own consistency! The extant solution to the puzzle with only two answers no longer applies.

The 3-valued puzzle introduced here might be seen as (at a least a bit) harder than all previous versions for at least three reasons: first, it is unexpected–the solution is not technically difficult, but only if we deeply understand how the logic adopted by the gods works. Second, it is in principle harder than the current hardest version by Uzquiano, where Verus and Falsus are granted the right to remain silent, while Aleatorius "... will gladly answer yes or no to the question of whether he and True or False would give the same answer to the question of whether Dushanbe is in Kirghizia." (Uzquiano 2010, p. 40). Here, however, the highly clever two-question solution of Uzquiano does not hold, since Aleatorius (or Random) may also reply 1/2. Our version amounts to Aleatorius being silent as well.

Thirdly, the amended version of Rabern and Rabern (2008) reduces to the present one by supposing *every* question to be consistent, that is, by taking $\circ(A)$ to be true for every statement, or in other words by responding with 'true' or 'false' (by means of two unknown words) to every question. Indeed, by adding $\circ(A)$ as an axiom schema, the three-valued logic **LFI1** reduces to classical propositional logic: since $\circ(A)$ is defined as $\neg\bullet A$, by clause (6) of the binary semantics described above, $I \not\models \bullet A$ for any interpretation I , and by clause (5) $I \not\models A \wedge \neg A$, thus by clause (1) $I \not\models A$ or $I \not\models \neg A$. Consequently, if $I \models A$ then $I \not\models \neg A$, reestablishing the full semantic characterization of classical propositional logic. The road, however, is open to make the 'hardest logic puzzle ever' as hard as you wish by playing with its internal logic. And this, far from being an exercise in futility, is an exercise in understanding logic and the scope of reasoning.

The ideas advanced here may encourage harder and harder logic puzzles, urging us to abandon any references to the 'hardest puzzle ever'. For instance, suppose the gods reply, always using the unknown adverbial sentences *da*, *ja*, and *ta*, but now based upon Lukasiewicz's three-valued logic **L3**. This logic has the same tables as the Kleene's strong logic of indeterminacy, differing just in the definition of implication. As is well known, the law of excluded middle and the law of non-contradiction are not tautologies in **L3**. However, Lukasiewicz logics (of any number of truth-values) are explosive since $A \to (\neg A \to B)$ are tautologies, and therefore are not paraconsistent. By supposing the gods remain silent on a question that they cannot reply without deductive explosion, we arrive at a quite difficult puzzle, whose solution I do not know.

Although the solution to the puzzle introduced here may be mathematically comparable to others, in a certain sense it is harder because it is surprising. We have to master new concepts, such as the idea that not all contradictions are equivalent, and that only consistent contradictions are dangerous, ideas unperceived in traditional logic. Rather than entering a competition to find shorter solutions or to produce the ultimately hardest form of a standard puzzle, my intention is to inaugurate not-yet-navigated forms of logic puzzles. It is certain that puzzles based on many-valued modal logics, intuitionistic logic, Belnap-Dunn's logic, Kleene's logic, or Lukasiewicz's infinite-valued logic will build harder and harder puzzles. It's about time that logic puzzle-makers take into account that logic is not reduced to what is called "classical logic", and try to actually use logic without the bridle of "classism". Puzzles will be even more fun, and I am sure Ray Smullyan would approve the new wave. To him this paper is dedicated.

Acknowledgements I acknowledge support from FAPESP Thematic Project LogCons 2010/51038-0, Brazil, from the National Council for Scientific and Technological Development (CNPq), Brazil, and from the Centre for Research on Architecture of Information, University of Brasilia. I am thankful to the two referees for their sharp comments and criticisms. Special thanks to Karen Kletter for her invaluable help in editing this paper.

References

Boolos, G. (1996). The hardest logic puzzle ever. *The Harvard Review of Philosophy*, *6*, 62–65.

Carnielli, W. A., & Coniglio, M. E. (2016). In *Paraconsistent Logic: Consistency, Contradiction and Negation*. Forthcoming in the series Logic, Epistemology, and the Unity of Science. Springer.

Carnielli, W. A., Coniglio, M. E., & Marcos, J. (2007). Logics of formal inconsistency. In D. Gabbay & F. Guenthner (Eds.), *Handbook of Philosophical Logic* (Vol. 14, pp. 1–93)., pp Amsterdam: Springer.

Carnielli, W. A., Marcos, J., & de Amo, S. (2000). Formal inconsistency and evolutionary databases. *Logic and Logical Philosophy*, *8*, 115–152.

Liu, F., & Wang, Y. (2012). Reasoning about agent types and the hardest logic puzzle ever. *Minds and Machines*, *23*, 123–161.

Novozhilov, N. (2012). The hardest logic puzzle ever becomes even tougher. http://arxiv.org/abs/1206.1926v1.

Rauser, R. (2002). Is the trinity a true contradiction? *Quodlibet Journal*, 4(4). Published online: http://www.quodlibet.net/articles/rauser-trinity.shtml.

Rabern, B., & Rabern, L. (2008). A simple solution to the hardest logic puzzle ever. *Analysis*, *68*, 105–112.

Smullyan, R. (1978). *What is the name of this book?* The Riddle of Dracula and Other Logical Puzzles. (New Jersey) Prentice-Hall, Inc.

Uzquiano, G. (2010). How to solve the hardest logic puzzle ever in two questions. *Analysis*, *70*, 39–44.

Wheeler, G., & Barahona, P. (2012). Why the hardest logic puzzle ever cannot be solved in less than three questions. *Journal of Philosophical Logic*, *41*, 493–503.

Wintein, S. (2011). A framework for riddles about truth that do not involve self-reference. *Studia Logica*, *98*(22), 445–482.

Wintein, S. (2012). On the behavior of True and False. *Minds and Machines*, *22*, 1–24.

Chapter 12
Bibliography of Raymond Smullyan

Raymond M. Smullyan

12.1 Raymond Smullyan's Books and Papers

Technical Books

Smullyan, R. M. (1961d). *Theory of formal systems* (Rev ed.). Princeton, NJ: Princeton University Press.
Smullyan, R. M. (1968b). *First-order logic* (Rev. ed., Dover Press, New York, 1994). Berlin: Springer.
Smullyan, R. M. (1992a). *Gödel's incompleteness theorems*. Oxford University Press.
Smullyan, R. M. (1993). *Recursion theory for metamathematics*. Oxford University Press.
Smullyan, R. M. (1994a). *Diagonalization and self-reference*. Oxford University Press.
Smullyan, R. M. (2014a). *A beginner's guide to mathematical logic*. Dover.
Smullyan, R. M. (2016a). *A beginner's further guide to mathematical logic*. World Scientific.
Smullyan, R. M., & Fitting, M. C. (1996). *Set theory and the continuum problem* (Rev. ed., Dover Publications, 2010. Errata at http://melvinfitting.org/errata/errata.html). Oxford University Press.

Technical Papers

Nerode, A., & Smullyan, R. M. (1962). Review of the foundations of mathematics, a study in the philosophy of science by Evert W. Beth. *Journal of Symbolic Logic, 27*(1), 73–75.
Smullyan, R. M. (1956a). Elementary formal systems (abstract). *Bulletin of the American Mathematical Society, 62*, 600.
Smullyan, R. M. (1956b). Languages in which self reference is possible (abstract). *Journal of Symbolic Logic, 21*(2), 221.
Smullyan, R. M. (1956c). On definability by recursion (abstract). *Bulletin of the American Mathematical Society, 62*, 601.

M. Fitting and B. Rayman (eds.), *Raymond Smullyan on Self Reference*,
Outstanding Contributions to Logic 14, https://doi.org/10.1007/978-3-319-68732-2_12

Smullyan, R. M. (1957). Languages in which self reference is possible. *Journal of Symbolic Logic, 22*(1) (Reprinted in *The Philosophy of Mathematics*, Jaakko Hintikka editor, Oxford University Press, 1969), 55–67.
Smullyan, R. M. (1958a). Theories with effectively inseparable nuclei (abstract). *Journal of Symbolic Logic, 23*(4), 458.
Smullyan, R. M. (1958b). Undecidability and recursive inseparability. *Zeitschrift für mathematische Logik und Grundlagen der Mathematik, 4*, 143–147.
Smullyan, R. M. (1959a). Creative sets (abstract). *Journal of Symbolic Logic, 24*(4), 315.
Smullyan, R. M. (1959b). Effective inseparability (abstract). *Journal of Symbolic Logic, 24*(4), 315.
Smullyan, R. M. (1959c). Further results on effective inseparability. *Journal of Symbolic Logic, 24*(4), 315–316.
Smullyan, R. M. (1959d). Review of methods of logic by Willard Van Orman Quine. *Journal of Symbolic Logic, 24*(3), 219–220.
Smullyan, R. M. (1960). Theories with effectively inseparable nuclei. *Zeitschrift für mathematische Logik und Grundlagen der Mathematik, 6*, 219–224.
Smullyan, R. M. (1961a). Elementary formal systems. *Journal of the Mathematical Society of Japan, 13*(1), 38–44.
Smullyan, R. M. (1961b). Extended canonical systems. *Proceedings of the American Mathematical Society, 12*(3), 440–442.
Smullyan, R. M. (1961c). Monadic elementary formal systems. *Zeitschrift für mathematische Logik und Grundlagen der Mathematik, 7*(2), 81–83.
Smullyan, R. M. (1962). On Post's canonical systems. *Journal of Symbolic Logic, 27*(1), 55–57.
Smullyan, R. M. (1963a). A unifying principle in quantiffcation theory. *Proceedings of the National Academy of Sciences, 49*(6), 828–832.
Smullyan, R. M. (1963b). Creativity and effective inseparability. *Transactions of the American Mathematical Society, 109*(1), 135–145.
Smullyan, R. M. (1963c). Pseudo-uniform reducibility. *Journal of the Mathematical Society of Japan, 15*(2), 129–133.
Smullyan, R. M. (1964). Effectively simple sets. *Proceedings of the American Mathematical Society, 15*(6), 893–895.
Smullyan, R. M. (1965a). A unifying principle in quantiffcation theory. In J. W. Addison, L. Henkin, & A. Tarski (Eds.), *The theory of models* (pp. 433–434), 1963 International Symposium on the Theory of Models. Berkeley: North-Holland Publishing Co.
Smullyan, R. M. (1965b). Analytic natural deduction. *Journal of Symbolic Logic, 30*(2), 123–139.
Smullyan, R. M. (1965c). On transfinite recursion. *Transactions of the New York Academy of Sciences II, 28*(2), 175–185.
Smullyan, R. M. (1966a). Finite nest structures and propositional logic. *Journal of Symbolic Logic, 31*(3), 322–324.
Smullyan, R. M. (1966b). Trees and nest structures. *Journal of Symbolic Logic, 31*(3), 303–321.
Smullyan, R. M. (1967a). More on transfinite recursion. *Transactions of the New York Academy of Sciences II, 29*(5), 555–559.
Smullyan, R. M. (1967b). Review of undecidable and creative theories by J. R. Shoenfield. *Journal of Symbolic Logic, 32*(1), 123.
Smullyan, R. M. (1967c). The Continuum Problem. In P. Edwards (Ed.), *The encyclopedia of philosophy* (pp. 207–212). New York: Macmillan.
Smullyan, R. M. (1968a). Analytic cut. *Journal of Symbolic Logic, 33*(4), 560–564.
Smullyan, R. M. (1968c). Uniform Gentzen systems. *Journal of Symbolic Logic, 33*(4), 549–559.
Smullyan, R. M. (1969). The continuum hypothesis. In A. George Boehm (Ed.), *The mathematical sciences a collection of essays*. Cambridge: MIT Press.
Smullyan, R. M. (1970). Abstract quantification theory. In J. Myhill, A. Kino, & R. Vesley (Eds.), *Intuitionism and proof theory, proceedings of the summer conference at Buffalo N. Y. 1968* (pp. 79–91). North-Holland.

Smullyan, R. M. (1972). Review of mathematics of incompleteness and undecidability by Vladeta Vučkovicć. *Journal of Symbolic Logic, 37*(1), 195–196.
Smullyan, R. M. (1973). A Generalization of intuitionistic and modal logics. In H. Leblanc (Ed.), *Truth, syntax and modality* (pp. 274–293). Amsterdam: North-Holland.
Smullyan, R. M. (1979b). Trees and ball games. *Annals of the New York Academy of Sciences, 321*, 86–90.
Smullyan, R. M. (1984b). Chameleonic languages. *Synthése, 60*(2), 201–224.
Smullyan, R. M. (1984c). Fixed points and self-reference. *International Journal of Mathematics and Mathematical Sciences, 7*(2), 283–289.
Smullyan, R. M. (1985a). Modality and self-reference. In *Intensional mathematics. Studies in logic and the foundations of mathematics* (Vol. 113, pp. 191–211). Amsterdam: North-Holland.
Smullyan, R. M. (1985b). Some principles related to Löb's theorem. In *Intensional mathematics. Studies in logic and the foundations of mathematics* (Vol. 113, pp. 213–229). Amsterdam: North-Holland.
Smullyan, R. M. (1985e). Uniform self-reference. In *Studia logica XLIV.4*. Tarski Symposium Volume, pp. 439–445.
Smullyan, R. M. (1986). Logicians who reason about themselves. In *Theoretical aspects of reasoning about knowledge. Proceedings of the 1986 conference on reasoning about knowledge* (pp. 341–352). Morgan Kaufman.
Smullyan, R. M. (1987b). Quotation and self-reference. In S. J. Bartlett & P. Suber (Eds.), *Self-reference* (pp. 123–144). Dordrecht: Martinus Nijhoff Publishers.
Smullyan, R. M. (1988). Review of mind tools. The five levels of mathematical reality, by Rudy Rucker. *Journal of Symbolic Logic, 53*(4), 1254–1255.
Smullyan, R. M. (1989). Logic puzzles. *Atti Degli Incontri di Logica Matematica, 5*, 73–83.
Smullyan, R. M. (1990). Some new double induction and superinduction principles. *Studia Logica, 49*(1), 23–30.
Smullyan, R. M. (1991a). Quantification theory in a nutshell. In J. H. Ewing & F. W. Gehring (Eds.), *Paul Halmos, celebrating 50 years of mathematics* (pp. 297–304). New York: Springer.
Smullyan, R. M. (1991b). Some unifying fixed point principles. *Studia Logica, 50*(1), 129–141.
Smullyan, R. M. (1996). Gödel's incompleteness theorems. In *Grolier encyclopedia of science and technology*.
Smullyan, R. M. (2001a). Equivalence relations and groups. In C. A. Anderson & M. Zeleny (Eds.), *Logic, meaning and computation: Memorial volume for Alonzo church* (pp. 261–271). Netherlands: Kluwer Academic Publishers.
Smullyan, R. M. (2001b). Gödel's incompleteness theorems. In L. Goble (Ed.), *The Blackwell guide to philosophical logic* (pp. 72–89). Hoboken: Wiley Blackwell.
Smullyan, R. M. (2004a). An isomorphism related to Gödel's fundamental operations. *Logic Journal of IGPL, 12*(6), 439–445.
Smullyan, R. M. (2006a). Self-reference in all its glory. In T. Bolander, V. F. Hendricks, & S. A. Pedersen (Eds.), *Self-reference. CSLI Lecture Notes 8* (Vol. 178, pp. 151–163). CSLI Publications.
Smullyan, R. M. (2006b). A logical miscellany. In R. E. Auxier & L. E. Hahn (Eds.), *The philosophy of Jaakko Hintikka. The library of living philosophers* (pp. 437–446). Open Court.
Smullyan, R. M. (2011). Duplex diagonalization. *Logic without frontiers, Festschrift for Walter Alexandre Carnielli on the occasion of his 60th birthday* (pp. 1–16). London: College Publications.
Smullyan, R. M., & Putnam, H. (1960). Exact separation of recursively enumerable sets within theories. *Proceedings of the American Mathematical Society, 11*(4), 574–577.

Popular Books

Smullyan, R. M. (1977). *The Tao is silent*. Harper and Row.

Smullyan, R. M. (1978b). *What is the name of this book?* Prentice-Hall.
Smullyan, Raymond M. (1979a). *The chess mysteries of Sherlock Holmes*. Alfred A: Knopf.
Smullyan, R. M. (1980). *This book needs no title*. Prentice-Hall.
Smullyan, R. M. (1981d). *The chess mysteries of the Arabian knights*. Alfred A: Knopf.
Smullyan, R. M. (1982a). *Alice in puzzle-land*. William Morrow & Co.
Smullyan, R. M. (1982c). *The lady or the Tiger?*. Alfred A: Knopf.
Smullyan, R. M. (1983a). *5000 B. C. — and other philosophical fantasies*. St. Martin's Press.
Smullyan, R. M. (1985d). *To mock a mockingbird*. Alfred A: Knopf.
Smullyan, R. M. (1987a). *Forever undecided*. Alfred A: Knopf.
Smullyan, R. M. (1992b). *Satan, Cantor and Infinity*. Alfred A: Knopf.
Smullyan, R. M. (1997). *The riddle of Scheherazade and other amazing puzzles, ancient and modern*. Alfred A: Knopf.
Smullyan, R. M. (2002b). *Some interesting memories: A paradoxical life*. Davenport, Iowa: Thinkers' Press Inc.
Smullyan, R. M. (2003). *Who knows?: A study of religious consciousness*. Indiana University Press.
Smullyan, R. M. (2007). *The magic garden of George B and other logic puzzles*. World Scientific.
Smullyan, R. M. (2009a). *A spiritual journey*. Praxis International Inc.
Smullyan, R. M. (2009b). *Logical labyrinths*. Peters: A. K.
Smullyan, R. M. (2009c). *Rambles through my library*. Praxis International Inc.
Smullyan, R. M. (2010). *King Arthur in search of his Dog & other curious puzzles*. Dover.
Smullyan, R. M. (2013a). *The Gödelian puzzle book*. Dover.
Smullyan, R. M. (2015). *Reflections*. World Scientific.
Smullyan, R. M. (2016b). *A mixed bag: Jokes, puzzles, riddles and memorabilia*. Sagging Meniscus Press.
Smullyan, R. M., & Bispham, P. (Eds.). (2009). *In their own words: Pianists of the piano society, popular articles*. Lulu Press.

Popular Articles

Smullyan, R. M. (1975). Simplicus and the tree, an open air symposium. In *University of Chicago magazine* (pp. 32–34).
Smullyan, R. M. (1978a). Puzzles. *The Mathematical Intelligencer*, *1*(2), 110–111.
Smullyan, R. M. (1982b). Review of logic matters, by P.T. Geach. *The Mathematical Intelligencer*, *4*(1), 39–40.
Smullyan, R. M. (1983b). The abduction of Annabelle. *The Sciences*, *23*(5), 13–16.
Smullyan, R. M. (1983c). The lie detective. *The Sciences*, *23*(3), 9–12.
Smullyan, R. M. (1983d). When I was a boy. *The Sciences*, *23*(4), 11–13.
Smullyan, R. M. (1984a). A Plague of lies. *The Sciences*, *24*(3), 67–69.
Smullyan, R. M. (1984d). How Kazir won his wife. *The Sciences*, *24*(1), 67–69.
Smullyan, R. M. (1984e). On the other hand. *The Sciences*, *24*(4), 62–63.
Smullyan, R. M. (1985c). Satan, Cantor and Infinity. *College Mathematics Journal*, *16*(2), 118–121.
Smullyan, R. M. (1994b). Anger, retribution and violence. In A. Gandhi (Ed.), *World without violence, can Gandhi's vision become reality?* (pp. 210–211). Limited: Wiley Eastern.
Smullyan, R. M. (1995a). Stereo viewers. In *Stereoscopy*.
Smullyan, R. M. (1995b). Three stereo puzzles by Raymond Smullyan. *Stereo World Magazine*, 22.
Smullyan, R. M. (1999). A curious paradox. In E. R. Berlekamp & T. Rogers (Eds.), *The mathemagician and pied puzzler, a collection in tribute to Martin Gardner* (pp. 189–190). Natick: A. K. Peters.

Smullyan, R. M. (2002a). Some tricks and paradoxes. In D. Wolfe & T. Rogers (Eds.), *Puzzlers' tribute, a feast for the mind* (pp. 341–348). Natick: A. K. Peters.

Smullyan, R. M. (2004b). Gödelian puzzles. In B. Cipra, et al. (Eds.), *Tribute to a mathemagician* (pp. 49–54). Wellesley: A. K. Peters.

Smullyan, R. M. (2008). Memories and inconsistencies. In E. D. Demaine, M. L. Demaine, & T. Rodgers (Eds.), *A lifetime of puzzles* (pp. 245–252). Boca Raton: CRC Press.

Smullyan, R. M. (2013b). Truth and provability. *The Mathematical Intelligencer*, *35*(1), 21–24.

Smullyan, R. M. (2014b). Afterword: Rambles and reflections. In J. Rosenhouse (Ed.), *Four lives, a celebration of Raymond Smullyan* (pp. 125–137). Mineola: Dover.

Smullyan, R. M. (2016c). Mineola. In J. Beineke & J. Rosenhouse (Eds.), *The mathematics of various entertaining subjects: Research in recreational math* (p. 7). Princeton: Princeton University Press.

CPSIA information can be obtained
at www.ICGtesting.com
Printed in the USA
LVHW06*1916280318
571483LV00001B/13/P